ROUTLEDGE LIBRARY EDITIONS:
GEOLOGY

Volume 15

GEOMORPHOLOGY AND TIME

GEOMORPHOLOGY AND TIME

J.B. THORNES AND D. BRUNSDEN

Routledge
Taylor & Francis Group
LONDON AND NEW YORK

First published in 1977 by Methuen & Co Ltd

This edition first published in 2020
by Routledge
4 Park Square, Milton Park, Abingdon, Oxon OX14 4RN
605 Third Avenue, New York, NY 10017

Routledge is an imprint of the Taylor & Francis Group, an informa business

British Library Cataloguing in Publication Data
A catalogue record for this book is available from the British Library

ISBN: 978-0-367-18559-6 (Set)
ISBN: 978-0-429-19681-2 (Set) (ebk)
ISBN: 978-0-367-22005-1 (Volume 15) (hbk)
ISBN: 978-0-429-27028-4 (Volume 15) (ebk)

J.B. Thornes and D. Brunsden

Geomorphology and Time

Methuen & Co Ltd

First published 1977 by Methuen & Co Ltd
11 New Fetter Lane, London EC4P 4EE

Typeset in Great Britain by
Preface Graphics Limited,
Salisbury, Wilts
and printed in Great Britain at the
University Printing House, Cambridge

ISBN 0 416 80080 7 *hardback*
ISBN 0 416 80090 4 *paperback*

To Rosemary and Elizabeth

Errata

p. xvi	*Quarterly Journal of the English Geological Society* should read *Quarterly Journal of Engineering Geology*
p. 136, line 30	respects *should read* represents
p. 137, line 10	variable. We *should read* variable, we
p. 142, line 20	M_x *should read* M_L
p. 144, (7.20)	$v_x = \frac{K\partial\phi}{\partial x}$ *should read* $v_x = -K\frac{\partial\phi}{\partial x}$
p. 144, (7.21)	$v_x = \frac{-K\partial h}{\partial x}$ *should read* $v_x = -K\frac{\partial h}{\partial x}$
p. 147, line 10	$(1-\mu)$ *should read* $(\mu - 1)$
p. 148, line 2	$\partial T/\partial t = K[\partial^2 T/\partial y^2]$ *should be deleted*
p. 149, line 23	care *should read* car
p. 152, line 8	$\partial z/(c)t$ *should read* $\partial z/\partial t$
p. 152, line 10	$\partial z/\partial t$ *should read* (c)
p. 152, 6 from bottom	$(x/x1)$ *should read* (x/x_1)
p. 160, line 2	$S = KA - z$ *should read* $S = KA^{-z}$
p. 179, fig. 9.2	*the caption should continue* The profiles are measured at different locations for which the date of cessation of basal erosion is known.
Index	53 *should read* 52, 54 *should read* 53, 55 *should read* 54

It is regretted that there are typographical errors which remain uncorrected in this edition through no fault of the editors.

Contents

General editors' preface viii
List of figures ix
List of tables xiv
Preface xv
Acknowledgements xvi

1 Introduction 1
2 The measurement of time 28
3 The measurement of variables in time 53
4 Analysis of temporal data 68
5 Rates of operation of geomorphological processes 93
6 Qualitative temporal models 112
7 Quantitative deterministic models of temporal change 133
8 Stochastic models of form evolution 155
9 Space and time 174

Bibliography 186
Appendix: The metric system: conversion factors and symbols 200
Index 203

The Field of Geography

Progress in modern geography has brought rapid changes in course work. At the same time the considerable increase in students at colleges and universities has brought a heavy and sometimes intolerable demand on library resources. The need for cheap textbooks introducing techniques, concepts and principles in the many divisions of the subject is growing and is likely to continue to do so. Much post-school teaching is hierarchical, treating the subject at progressively more specialized levels. This series provides textbooks to serve the hierarchy and to provide therefore for a variety of needs. In consequence some of the books may appear to overlap, treating in part of similar principles or problems, but at different levels of generalization. However, it is not our intention to produce a series of exclusive works, the collection of which will provide the reader with a 'complete geography', but rather to serve the needs of today's geography students who mostly require some common general basis together with a selection of specialized studies.

Between the 'old' and the 'new' geographies there is no clear division. There is instead a wide spectrum of ideas and opinions concerning the development of teaching in geography. We hope to show something of that spectrum in the series, but necessarily its existence must create differences of treatment as between authors. There is no general series view or theme. Each book is the product of its author's opinions and must stand on its own merits.

University of London
King's College

W. B. Morgan
J. C. Pugh

List of figures

1.1a	Hypothetical resultant and vectors of net sediment transport in an estuary	4
1.1b	Relationships between channel discharge, velocity and geometry at a station through time and in a downsteam direction	4
1.2	Observation of runoff at Iping Mill weir, river Rother, Sussex	6
1.3	Hypothetical distribution showing the frequency of events yielding particular amounts of debris at the head of a talus slope in one year	7
1.4	Hypothetical record of mudflow movement	8
1.5	Average number of frost cycles for a meteorological station for each month of the year (hypothetical)	8
1.6	Cumulative frequency of discharge expressed as accumulated percentage of time against magnitude of event	9
1.7	The return periods of discharge events on the Lycoming Creek, Pa.	9
1.8	Rainfall histogram and hydrograph showing the relationship between inputs and outputs for a single storm event on the Delaware river	10
1.9	A model for feedback in a coastal mudflow	11
1.10	Schematic diagram of a cascaded landslide system in Dorset	12
1.11	The spatial variability of mass movement between 1887 and 1960, Stonebarrow, Dorset	14
1.12	Dynamic response of an Antrim mudslide to precipitation after a threshold has been exceeded	16
1.13a	Position of slope profiles relative to a retreating ice front in Iceland showing changing basal conditions	17
1.13b	The successive states and relaxation path assumed by the talus slopes through time from retreat of the basal ice	17
1.14	Systems adjustments after 'shock' indicating the magnitudes of the relaxation times	18

1.15a Histograms of slope angles on Pennine grits and Devonian slates 21
1.15b Deductive model of slope evolution developed by Young (1963) to illustrate the development of accumulation features as a consequence of downslope soil transport 21
1.15c Deductive models of slope evolution developed by Young (1963) for three sets of model assumptions 21
1.15d Slope profiles showing suggested change of form following the retreat of the sea from the base of a slope in South Wales 21
1.16 The development of Hurst Castle Spit simulated by computer 25
2.1 Schematic block diagram showing the final stage of the geological history of a Tripolitanian wadi 29
2.2 Examples of isochronous and metachronous surfaces produced by different schemes of landscape evolution 31
2.3 Simplified pollen diagram from landslide sites at Coombes Rocks 33
2.4 The alluvial fill of the Ouse valley, Sussex 34
2.5 Averaged climatic curves shown by ocean cores in the north and equatorial Atlantic and the Caribbean 35
2.6 A molluscan fauna diagram with radio-carbon date for Dover Hill, Folkestone 37
2.7a Changes in $CaCo_3$ with depth for a buried soil in Wyoming 38
2.7b Changes in soil properties with depth for buried soil layers 38
2.8 The ash deposit from the Quizapu volcano in Chile which erupted in 1932 40
2.9 Dendrochronology 45
2.10 Recessional sequence of the Emmons Glacier based on dendrochronology 46
2.11 Chronology for stages of retreat of continental ice from Sweden based on de Geer's analysis of varves 47
2.12 Regional growth rate curves for *Rhizocarpon geographicum* for different Scandinavian locations 50
2.13 The use of *Rhizocarpon geographicum* in the Tunsbergdal valley, Norway, for dating moraine sequences 51
3.1 The Landslip Quadrille 53
3.2 Changes in spit outline on Dawlish Warren at the mouth of the river Exe 54
3.3 Air photographs of Black Ven landslide in 1958 and 1969 to illustrate the increasing importance of repetitive cover in historical geomorphological studies and surveys of rates of erosion 55
3.4a Discharge record for the river Mecham, Queen Elizabeth Islands, 1970 60
3.4b Cumulative movement on three mudflows in Dorset 60
3.4c Soil temperatures at Resolute Bay 60
3.5 Sampling schemes for various types of temporal data 61
3.6 Example of a clockwork mechanism for use in obtaining a continuous trace of frost heave in soils 63

3.7	The tipping bucket mechanism to record event occurence	64
3.8a	Variation in computed information content with sampling interval	67
3.8b	Variation in smoothed information content with sampling interval	67
4.1	Log-log regression of independently observed random variables	69
4.2	Methods of depicting frequency count information	71
4.3	Effects of changing average value on the Poisson distribution	73
4.4	Cumulative exponential distribution for the probability that time between arrivals of avalanches exceeds *t* units of time when mean arrival rate is one per week	74
4.5	Recurrence interval in years for the Youghiogheny and Kiskiminetas rivers	76
4.6	Gumbel distribution of annual maximum 30 min precipitation partial series	78
4.7a	Fluctuations in mean annual nitrate concentrations for the river Stour 1937-71	79
4.7b	Sea level change in the last 9,000 years	79
4.8	Hypothetical response of sediment yield to devegetation	80
4.9a	Discharge of the river Stour at Langham, Essex	81
4.9b	Notation for the cosine curve	81
4.10	Variance spectrum for the discharge of the river Stour at Langham	83
4.11	Hypothetical correlogram for height of channel bed	85
4.12a	Different averaging schemes for sampled data	86
4.12b	Scale effects of different averaging procedures	86
4.13	Weekly fluctuations in discharge and nitrate for the river Stour at Langham	88
4.14	Coherence between dissolved solid load and discharge for the Animas river in Texas	89
4.15	Hypothetical diagram of cascaded input-output systems involving thresholds	90
4.16	Application of the unit hydrograph method	91
5.1	Example of ordinal and interval scales for classifying mass movement	99
5.2	Examples of instruments which might be used to measure throughflow and soil wash in the field	104
5.3	The use of erosion pins to measure the rate of degradation of a rapidly evolving clay slope	105
5.4	An example of an attempt to measure overall denudation rates from erosional evidence where volume removed in a known period of time is calculated from successive time levels of an erosion surface	107
5.5	The relationship between precipitation amount and intensity and sediment yield derived by Fournier to distinguish different climatic regions	109

6.1 Diagram of a typical 'black-box' or transfer function approach to geomorphology 115
6.2 Cyclic models of landscape evolution showing the relationship between elevation and time 122
6.3 Part of the regional model for the London Basin developed by Wooldridge and Linton (1955) 124
6.4 Graphs developed by Schumm to examine the effect of climate on landform and sediment yield 126
6.5 The record of sediment, processes and climatic inferences discussed by Butzer (1964) for the regional chronologies of Torralba and Ambrona, S. Spain 128
6.6 Detailed map of a landslide complex at Stonebarrow Hill, Dorset 130
6.7 General model suggested by Price (1969) for a complex pro-glacial environment 132
7.1 Graph to show the derivation for a tangent to a curve 134
7.2 Tangential plane to a surface at a point P 136
7.3 Terminology for the derivation of slope evolution equations 137
7.4 Decay of a vertical cliff where the change in height with time is a function of curvature 139
7.5 Hypothetical infiltration response to precipitation 141
7.6 The damping of groundwater inflow and outflow to a steady state condition caused by the passage of a flood wave in a stream channel 142
7.7 Notation for development of steady state diffusion model 143
7.8 Cliff recession under the models of Bakker and le Heux (1952) 146
7.9 Terminology for formulation of the continuity equation 147
7.10 Decay of thermal effects of a positive increase in temperature at the surface of a glacier as the temperature wave is transmitted down through the ice 148
7.11 Decay of a mountain under the assumptions of the diffusion model for three successive time periods 149
7.12a Passage of a surface wave of ice down the Mer de Glace 151
7.12b Wave paths in the x, t plane for an idealized glacier 151
7.13 Solutions to the Kirkby (1971) model according to different values of the parameters m and n 153
7.14 The effect of perturbations on surfaces 153
8.1 Examples of simulated stream patterns using Monte Carlo techniques 160
8.2 Examples of random walks 163
8.3 Diagram of a 'probablistic' frog 164
9.1 The Cartesian coordinate system used in conventional three-dimensional notation 177
Sequential slope profiles for a Mississippi river bluff near Port Hudson, Louisiana 179
9.3 Three maps which represent successive time slices and show spatial changes in the position of the Mississippi river for 1883, 1921 and 1971 180

9.4 The passage of a point (sea-level) transverse to the coast, through space and time 181
9.5 Manifold in the space-time cube, with space in two dimensions, showing the hypothetical development of a meander 182
9.6 The space-time cube representing geographical data matrices 183

List of tables

1.1 Status of variables during designated time spans 13

1.2 Classic model of the denudation chronology of south-east England 20

3.1 Some common types of measurement in time 58

3.2 The movement of talus material in the Lake Louise area 59

4.1 The amount of daily melt at a temperate glacier snout 70

4.2 Maximum annual rainfall values 1940-61 for Madrid 77

5.1a Rates of uplift estimated for orogenic and isostatic conditions 94

5.1b Rates of denudation estimated for drainage basins 94

5.2 Rates of erosion under different climates for mountain and plain areas of the world 95

5.3 Example of the types of variable used to measure erosion rates 97

5.4 The relative importance of processes in a periglacial mountain climate 110

6.1 A classification of models based on their function 113

6.2 Examples for different qualitative geomorphological model formulations 116

9.1 Classification of types of model based on space-time structuring 176

Preface

Our purpose in writing this book is to examine time, one of the central themes of geomorphology, in an attempt to provide an overall framework within which to consider and compare old and new approaches to the subject. We include, therefore, in the discussion reviews of the qualitative model-building approach of the pre-1950s, the empirical, observation, process measurement and field measurement technique of the 1960s, and currently fashionable analytical modelling. We have tried to avoid a compilation of existing literature in textbook or manual style for we wanted to write about ideas and conceptual approaches. Many subjects are therefore only briefly mentioned, left unfinished or not fully explained but hopefully there is sufficient to implant an idea or stimulate thought, discussion and further reading. If the book achieves that aim we hope that the ensuing ferment will cause less cerebral damage than it inflicted on the authors.

We are most grateful to the many patient people who assisted in the production of the book. Mrs Anne Rogers and Miss Penny Roberts who typed the manuscript over and over again; the late Miss Duki Orsanic, Miss Roma Beaumont, and Mr Gordon Reynell who drew the figures; Miss Patricia Aylott who was responsible for the photographic work; and Professor J. C. Pugh who read and reconstructed the manuscript with a selfless thoroughness and sacrifice of time. Finally to Rosemary and Elizabeth who must be mad to put up with us.

Acknowledgements

The authors and publishers would like to thank the following for permission to reproduce copyright material:

West Sussex River Authority for fig. 1.2
D. R. Coates and Southern University of New York for fig. 1.7
John Wiley & Sons Inc. for data in fig. 1.8
American Journal of Science for table 1.1, fig. 5.4
Bulletin, Geological Society of America for figs. 1.12, 8.1
Institute of British Geographers for figs. 1.13, 1.15a, 1.15d, 2.3, 2.4, 3.2, 3.4a, 5.3, 6.7, 7.13
American Society of Civil Engineers for figs. 1.4a, 3.8
American Geophysical Union for fig. 1.4b
Royal Society *Philosophical Transactions* for figs. 1.11, 2.6, 7.12b
Journal of Geology for figs. 1.4c, 1.16, 7.4, 10.2
Anthony Young for fig. 1.5b
George Phillip for data in table 1.2 and fig. 6.3
Cambridge University Press for figs. 2.1, 7.9, table 5.1
Geological Society, London for fig. 2.5
W. H. Freeman & Co for figs. 2.7a, 4.5, 4.11a
Journal of Soil Science for fig. 2.7b
Uppsala Geological Institute for fig. 2.8
United States Geological Survey for figs. 2.9, 2.10, 7.6, 8.2b, table 5.1
Interscience, New York for fig. 2.11
Norsk Geografisk Tidsskrift for fig. 2.12
Geografiska Annaler for fig. 2.13, table 5.4
Royal Air Force and Fairey Aviation for fig. 3.3
Geographical Bulletin for fig. 3.4c
Zeitschrift fuer Geomorphologie for table 3.2
British Geomorphological Research Group for figs 3.6, 3.7, 5.2a-b
Min. of Agric. Servicio de Conservacion de Suelen, Madrid for tables 4.2, 4.6
Geol. en Mijnbouw for fig. 4.7b
Geographical Analysis for fig. 4.12b
Annales de Geographie for table 5.2
United States Highway Research Board for fig. 5.1
Edward Arnold Publishers for fig. 5.2c, table 6.1
Presses Universitaires de France for fig. 5.5
Macmillan, New York, for fig. 6.2b
Princeton University Press for fig. 6.4
Methuen & Co Ltd for fig. 6.5
Quarterly Journal of the English Geological Society for fig. 6.6
Springer-Verlag for figs 7.3, 7.11
Pergamon Press for figs. 7.10, 7.12a
Water Resources Research for fig. 1.14
Geographical Magazine for fig. 8.1b
United States Army Corps of Engineers (maps) for fig. 9.3
UNESCO for fig. 9.4
Association of American Geographers for fig. 9.6

1 Introduction

'The leading idea which is present in all our researches, and which accompanies every fresh observation, the sound which to the ear of the student of Nature seems continually echoed from every part of her works, is —

Time!——— Time! ——— Time!'

George Poulett Thomson Scrope 1858

Time pervades all fields of geomorphology, from the most restricted observation of an obscure localized geomorphological process, for example the solution of minerals from potholes in glacially eroded granites, to the macroscopic, highly abstract modelling of drainage systems development over several thousands of years. Moreover, some of the most important geomorphological concepts, are time-centred — equilibrium and grade, magnitude and frequency, equifinality, denudation chronology, variance minimization and entropy maximization.

The various interests range over the finite span of time of which we have knowledge: the development of erosion surfaces across the Archean basements; the sequence of uplift and erosion in the late Tertiary; the chronology of glacial and deglacial stages in the Quaternary; the lateral translation of fluvial activity on the flood plain; the passage of a peak of sediment in the channel or the variations in stress on a sand particle in the turbulence of a laboratory flume.

The techniques we use to observe the phenomenon under investigation are conditioned by the time scale and time characteristics of the processes or forms which are deemed important. There is, for example, little place for continuous observation of snow cover in temperate latitudes; the spacing of observations on wave height must take into account the periodic nature of the phenomenon both in time and space, and similarly observations of mudflow activity at very long intervals yield cumulative results which are probably of very limited use in illuminating the principal factors generating and controlling the flow, especially if the response rate to the factors controlling the flow is rapid.

The centrality of time in geomorphology stems from three fundamental developments in the study of the subject. These are, in order of development, (i) denudation chronology and evolutionary studies, (ii) accurate description of the mechanism and rates of operation of geomorphic processes, (iii) the adoption of a systems-based attitude towards geomorphological investigations. It is ironic, but by no means fortuitous, that time — the central theme of denudation chronology — plays so important a role in the latter studies. It seems to make good sense to rearrange the order of these issues to represent increasing time spans, decreasing time precision and a logical change in the observation, model-building and model-testing techniques. We also choose to progress from scientifically general ideas about time and observation in time to those which are particularly 'geomorphological'. The difficult subject of denudation chronology therefore comes late in our discussion. At the same time, it is not the purpose of this book to perpetuate the rather fruitless polarization of the subject between denudation chronology and contemporary processes studies, or between advocates of time dependency and time independency. Least of all are we able to differentiate between quantitative and non-quantitative. The problems of contemporary process studies are just as difficult as those of long-term evolution; the latter are just as capable of precise and systematic study and each should be complementary to the other.

This book is about *one* of the central issues of geomorphology, not *the* central issue. Lest it should be thought that it is exclusive, recall that other equally central issues exist — lithology and process, geomorphology and climate, spatial variability, or stochastic processes. All of these themes impinge on the problems of observation, model-building and interpretation of events in time.

Geomorphological data and temporal explanation

Before discussing the more important ways in which geomorphologists are concerned with time, it is useful to summarize a few important aspects of time in terms of the nature of geomorphological data and the common modes of temporal explanation.

Data

Many of the problems described in this book will be concerned with the relationship between time and space, and it is therefore important to recognize that they do not always possess analogous properties. The fundamental temporal attribute of duration may be compared to the spatial qualities of area or distance in that they are of finite and of measurable magnitude. In terms of temporal or spatial relationships of phenomena, however, time is distinguished by possessing the property of intrinsic direction and in the macroscopic sense being irreversible. Thus we speak of past, present and future, of being and becoming. Events begin, endure and end and are fixed in position, and therefore direction, relative to preceding and succeeding events. There is a striking difference with spatial phenomena which must be fixed in space by reference to at least three points. This transitory, directed nature of time is fundamental to any understanding of process where we seek to establish the rate of operation, direction, duration, memory of preceding events or relaxation times.

Geomorphological data may be classified on a time basis for they may possess several distinctive properties. For example, discrete events may be regarded as 'isolated' if they occur so infrequently that they may be regarded as of 'once only' character. More commonly, they possess the properties of *continuity*, such as the temperature of a rock mass; *fluctuation*, as the temperature rises and falls, and *sequential* occurrence as a *time-series*. In such a sequence it is possible to recognize simple patterns which may be regular, random or clustered; short or long term trends such as evolution and succession; cycles, or intermittent sequences such as might occur in volcanic events or movements along a fault.

Geomorphological phenomena may be regarded as slowly changing or dynamic, if measured against the human time scale. Slowly changing phenomena include sea-level change, isostatic rebound and the ground loss implied in the ideas of peneplanation or pediplanation, though in practice such changes are difficult to observe. In these ideas the concept of movement is a lesser consideration than in phenomena that are regarded as dynamic.

Velocity and acceleration are vital elements of geomorphological process studies. Velocity, expressed as the distance moved in unit time (L/T), is among the most common of all process measurements and is often combined with direction to yield vectors and resultants of position or system states (fig. 1.1a). The net direction and amount of sediment transport is a useful example. This also serves to illustrate the close relationships which exist between time and space, for we not only combine them to yield velocity (L/T) but we also compare the values in further time-space terms. Changes in stream velocity at one point in space (gauging station) or at several points in space (several gauges downstream) (fig. 1.1b) demonstrate the notion as do graphs of distance moved in unit time or studies of kinematic wave phenomena (Nye 1965).

Dynamic phenomena may also be regarded as random, unidirectional or cyclical in character. Interest in randomly-occurring phenomena is rapidly increasing as geomorphologists move away from deterministic to probabilistic modes of explanation. Recently, for example, Howard (1965) has suggested that landslides may be random events; Culling (1963) has attempted to develop a theory of soil creep based on a consideration of random forces in a soil mass. Most geomorphological phenomena, however, incorporate a strong unidirectional space-time element. Most commonly, this includes the influence of gravity which is the dominant force in all geomorphological processes. Cyclical characteristics often occur where climatic fluctuations are involved. The most notable examples come from fluvial geomorphology.

Finally, the results of geomorphological events may be regarded as

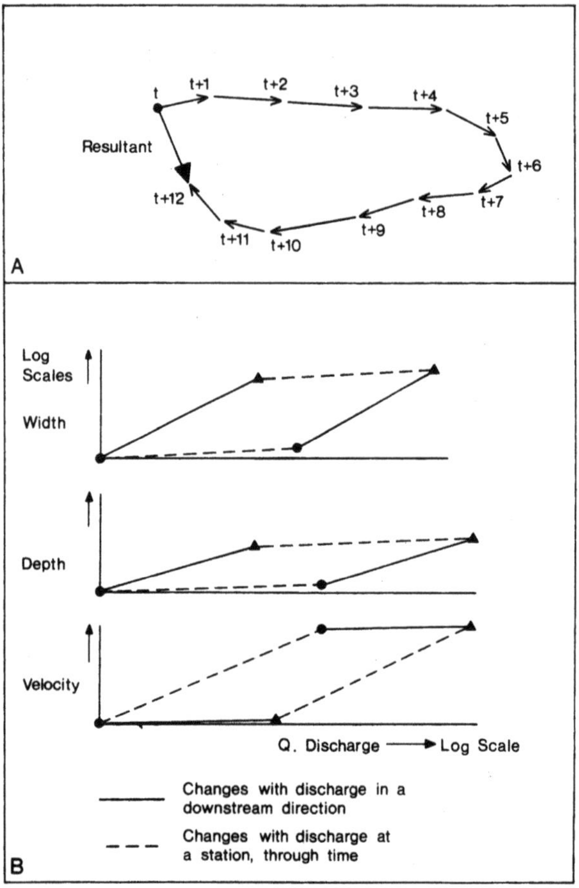

1.1 (a) Hypothetical resultant and vectors of net sediment transport in an estuary. Each arrow represents the successive paths of a particle. The resultant represents the net effect of these movements. (b) Relationships between channel discharge, velocity and geometry at a station through time and in a downstream direction.

reversible or irreversible. The former is implicit in ideas of steady state beach profiles where sediment is lost and returned. Total ground loss from a drainage basin or landslides from a cliff in the short term are examples of irreversible change of the landform geometry.

Explanation

It is always difficult to remove elements of subjectivity from our view of time since we are influenced by our own experience of natural phenomena (Meyerhoff 1960). There are real dangers here in our attempts at temporal explanation, for as with any model the temporal model must be verified by empirical statements which may not fulfil our subjective ideas.

Perhaps it is not surprising that many attempts to explain geomorphological phenomena had a genetic and evolutionary bias closely related to the experience of man. Thus, for example, Davis (1909) applied the human concepts of youth, maturity and old age to landscape development.

In considering these problems Harvey (1969) suggests that the most common types of temporal explanation in geography include (a) narrative, (b) reference to time or stage as an explanatory variable or (c) employment of some process mechanism either real or hypothesized.

Narrative, the historian's 'special tool', is often used by geomorphologists to describe the occurrence of events in time. The best accounts are purely descriptive but it is difficult to omit allusion to explanatory statements, subjective selection of 'significant' events or meaningful hints to associations, relationships or causality which are more imaginary than real. Narrative is not a rigorous or consistent method of temporal explanation.

Equally common in geomorphology are attempts to use time as one of several explanatory variables, e.g. 'landforms are a *function of* structure, process and *time*'. As Stoddart (1966) has shown the Darwinian concept of evolution was reinterpreted by Davis (1909) so that 'what for Darwin was a process became for Davis and others a history'. In this method of explanation it is normal to establish a time scale to fit the proposed sequence of events and to hypothesize about the mechanisms whereby a particular evolutionary sequence is achieved. Thus Davis proposed the cycle of erosion over cyclic time as an analogue to the passage of the life of an individual. When such a model is tested by recourse to empirical findings it is usual to find modifications creeping in. Davis' 'interruptions' to the geographical cycle were a necessary adjunct to an imperfect model.

The major cause of failure, however, was the lack of knowledge of the processes (in this case erosion and deposition) which were responsible for the proposed evolutionary stages. For this reason Harvey (1969) has suggested that a more logical approach would be to start with processes, either hypothesized or real, and relate these to a time scale, artificial or real, in any given (geomorphological) situation. He quotes Hutton who assumed that if 'the present was the clue to the past' then present processes could explain the evolution of events in the past.

Fortunately, current process studies in geomorphology are yielding potentially better explanations about process and we can look forward to explanations that are based on careful process response models such as those of Gilbert (1877), Schumm (1956) and Kirkby (1971). We seem now to be better informed than when Leighly wrote: 'Davis' great mistake was the assumption that we know the processes involved in the development of landforms. We don't and until we do we shall be ignorant of the general course of their development.' (Leighly 1940)

Description of events in time

From a purely observational point of view, as well as in classification, the ideas of frequency and rate are of paramount importance. For the purposes of description time can be considered under the headings: continuous,

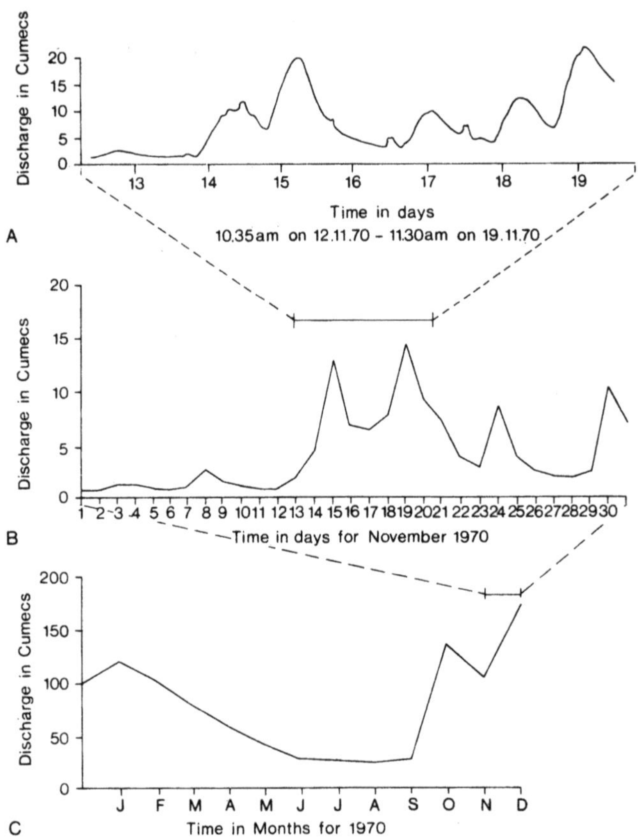

1.2 Observation of runoff at Iping Mill weir, river Rother, Sussex (courtesy of the West Sussex River Authority). (a) Continuous record. (b–c) Quantized record.

quantized, discrete or sampled time. *Continuous* means that observation is unceasing. Suppose we observe something which varies in magnitude (a variable) all the time and are able to record that variable as time passes, without a break, then we say the observations are continuous. Fig. 1.2a shows a record of discharge in a stream channel; this record is continuous in time. Alternatively, we may *quantize* time, i.e. divide continuous time into imaginary sections — for a week, or a year. Here we are using discrete units based on modules of 7 or 365. The discharge record in fig. 1.2b-c is of quantized time. In this diagram the discharges for a whole month and a year have been summed, the data being obtained by continuously summing in a given time unit. In a technical sense, any time we abstract data from a continuous record we are quantizing time. We may, for the month's record, add together the hourly, minute, second or tenth-second observation. The purpose of the study, the data available and logistic reasons usually determine the choice of unit. The choice is critical for it will determine the loss of information which we seek to minimize. Obviously the smallest unit which can be measured is still discrete, noted by Δt. The problem of making Δt very small, and yet summing over quantized time, is well known in mathematics, physics and engineering and is known as integration. This process is symbolized as $\int_0^{0.5}$ where the figures indicate the boundary over which integration takes place.

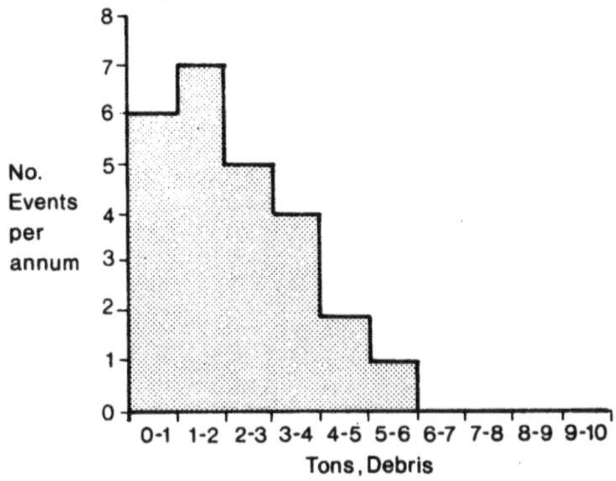

1.3 Hypothetical distribution showing the frequency of events yielding particular amounts of debris at the head of a talus slope in one year.

There are two other important ways of considering time in a descriptive sense. The first is to divide the time into packages where interest focuses on the length of the package, rather than the relationship between the packages in quantized time. This will be called *discrete* time. It is indicated by the terms *per* day, *per* year and so on, and is used to express the notion of frequency. Frequency is the number of events per discrete unit of time. For example (fig. 1.3) we could express the time component in the arrival of particles at the head of a talus slope by recording the weight in metric tons which arrives in a day, a week or a year, and then plotting the number of occasions on which 0, 1, 2, 3 tons arrived. The graph is hypothetical but we might expect that if our discrete time is one year, the number of occasions when 0 tons arrived was quite large, of one ton larger and then for 3, 4 and 5 tons the number of occasions would drop off until, as we might expect, the number of events when 6 tons were observed was very small.

The fourth and final way of observing, representing and interpreting from time is to think of *sampled* time (fig. 1.4). We imagine that a mudflow is in continuous flow, but we can only observe that flow for one day in every week. We then sample our variable (rate of flow) in a discrete block (one day) of quantized time from a continuous set of data (the whole period of time over which the mudflow was moving). Alternatively, on one day we observe how much movement has occurred in the preceding week. In this case sampled time is a quantized block of one week.

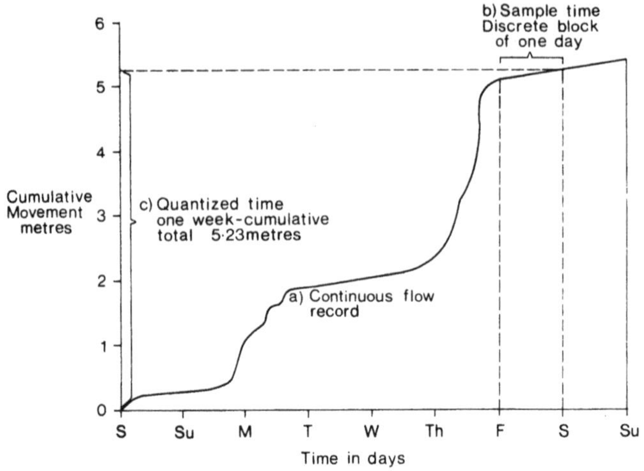

1.4 Hypothetical record of mudflow movement. (a) Continuous record. (b) The idea of sample time (one day) during which a sample movement of 0.25m took place. (c) Cumulative total movement for one week (quantized time).

Frequency has been expressed in terms of the number of events of a given size in a unit of time. Another expression of frequency is in terms of adjacent time units, in fact quantized time. Thus, we might indicate relative frequency in terms of months of the year, say the number of freeze-thaw cycles on average for each month throughout the year. This demonstrates how frequency itself changes from one time to another (fig. 1.5). Finally, we may express frequency in a cumulative sense, which could indicate the frequency with which a particular event or an event of lesser magnitude occurs. Thus, by

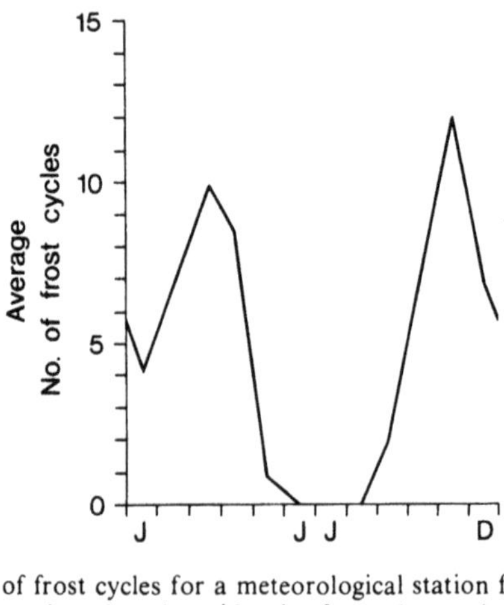

1.5 Average number of frost cycles for a meteorological station for each month of the year (hypothetical). Values are plotted at the mid-point for each month.

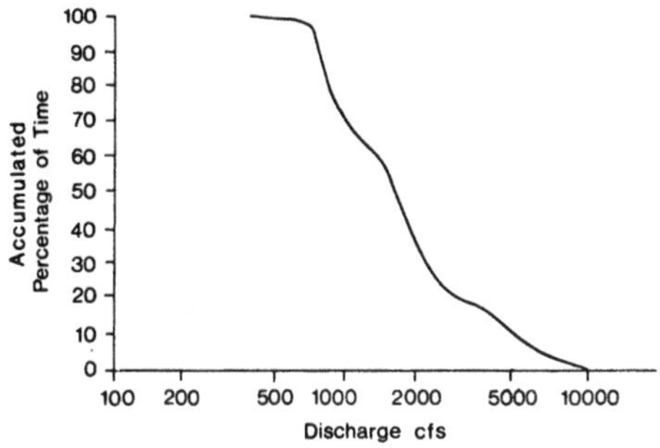

1.6 Cumulative frequency of discharge expressed as accumulated percentage of time against magnitude of event. The vertical axis gives the percentage of time for which the discharge is equalled or exceeded.

expressing cumulative frequency as percentage (fig. 1.6) we may observe, for example, what percentage of discharge events are equal to or less than a given magnitude.

The product of the relative frequency of an event and the magnitude of the event itself yields the graph of work done. This concept of combined effects of magnitude and frequency is an important one in geomorphology. It indicates that in many situations most of the work is effected by events of moderate intensity and frequency. This is because very large events do not occur frequently enough to do a great deal of work. This notion is illustrated in fig. 1.7, which shows the relative frequency and magnitude of discharge events on the Lycoming Creek, Pa. Note that since many phenomena have a Gaussian or bell-shaped frequency curve, one is often implying that 'moderate' means 'average' event.

For a given process there are spatial variations in its magnitude and frequency. For example, it is possible to compare the *relative* amount of

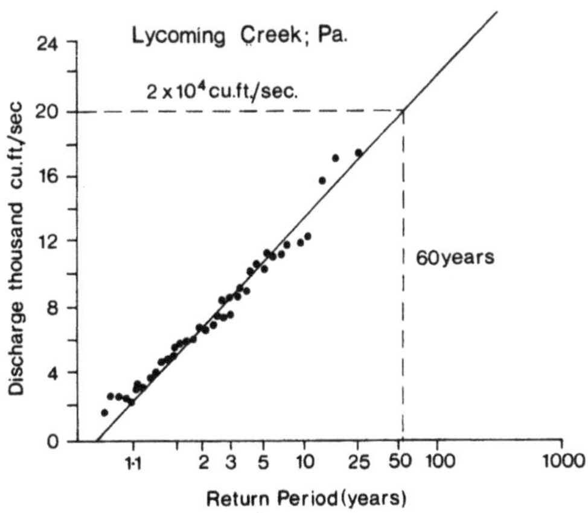

1.7 The return periods of discharge events on the Lycoming Creek, Pa. (Reiche 1971).

dissolved solids produced by rivers per annum in different parts of the world. Similarly, different processes do different amounts of work in the same area.

If we observe the concentration of dissolved solids in a particular time period as a continuous record, the resulting data are called a time series. The events in sampled time may be regarded as independent of each other and are often described by the characteristics (moments) of their frequency distributions. Special problems arise, however, for continuous and quantized time data because such observations are *not* usually independent of one another; although even this situation may occasionally occur. This theme is developed more fully in chapter 3.

Geomorphological systems and time

The systems concept is now widely employed in process studies and will be increasingly important in establishing the response to these processes. An extensive treatise on systems in physical geography, and particularly in geomorphology, is given in Chorley and Kennedy (1971). Basically, a system is a set of objects or attributes together with the relationships between them which are organized to fulfil a particular function. In terms of time, the systems of interest are those which convert a given input at time t to an output at time $t+k$. In this simple concept we regard the system as an 'operator' in the sense that it operates on the input to produce the output. There are, of course, many alternative ways of looking at the same system: Chorley and Kennedy (1971) also develop, for example, the process-response and correlation structure views of a system.

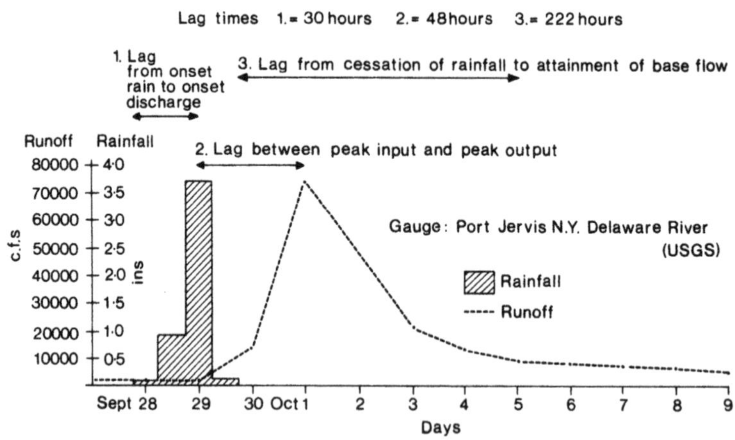

1.8 Rainfall histogram and hydrograph showing the relationship between inputs and outputs for a single storm event on the Delaware river (after Strahler 1969). By Day 5 the discharge is again comprised entirely of base flow, though this is at a higher level than base flow at the start of rainfall. Lag 3 is therefore from the cessation of rainfall to attainment of base flow since a common position for base flow for all storms is unknown.

For our purposes the most important point about natural systems is that they are non-anticipatory. This simply means that the system cannot forecast and adjust for a change in output before the output occurs; a situation which is common in human systems. Geomorphological systems are not able to adjust to things which have not yet happened — and so avoid them. This point is not trivial, since it means that by observing input and output we may learn more easily something of the way in which the system operates.

Input and output in systems may appear quite simple. The drainage basin may be regarded as a system in which the input is rain and the output is the discharge of the trunk stream and groundwater. The temporal relationships between them may be expressed in terms of the lag time between the onset of rain and the peak discharge for a particular type and size of flood (fig. 1.8). Another apparently simple input-output is that in which snow arriving at the head of a glacier in exceptional amounts is transmitted to the glacier snout. Such apparent simplicity is, however, complicated by many factors. One of these is the fact that systems may be able to store materials for varying lengths of time, another that feedback occurs in the system, sometimes damping the input effects, at other times exaggerating them (fig. 1.9).

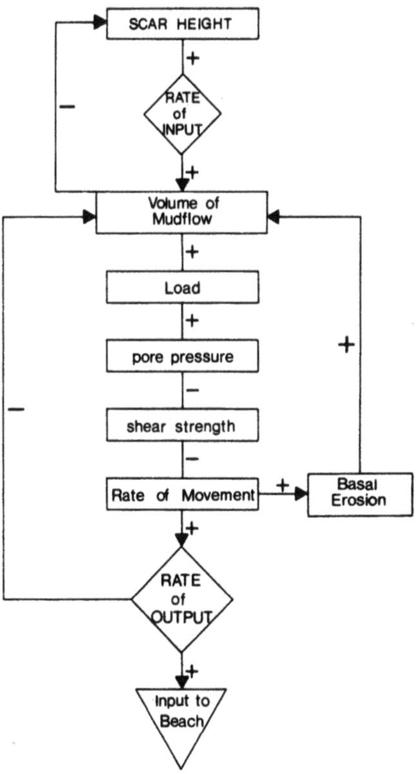

1.9 A model for feedback in a coastal mudflow (after Brunsden 1973).

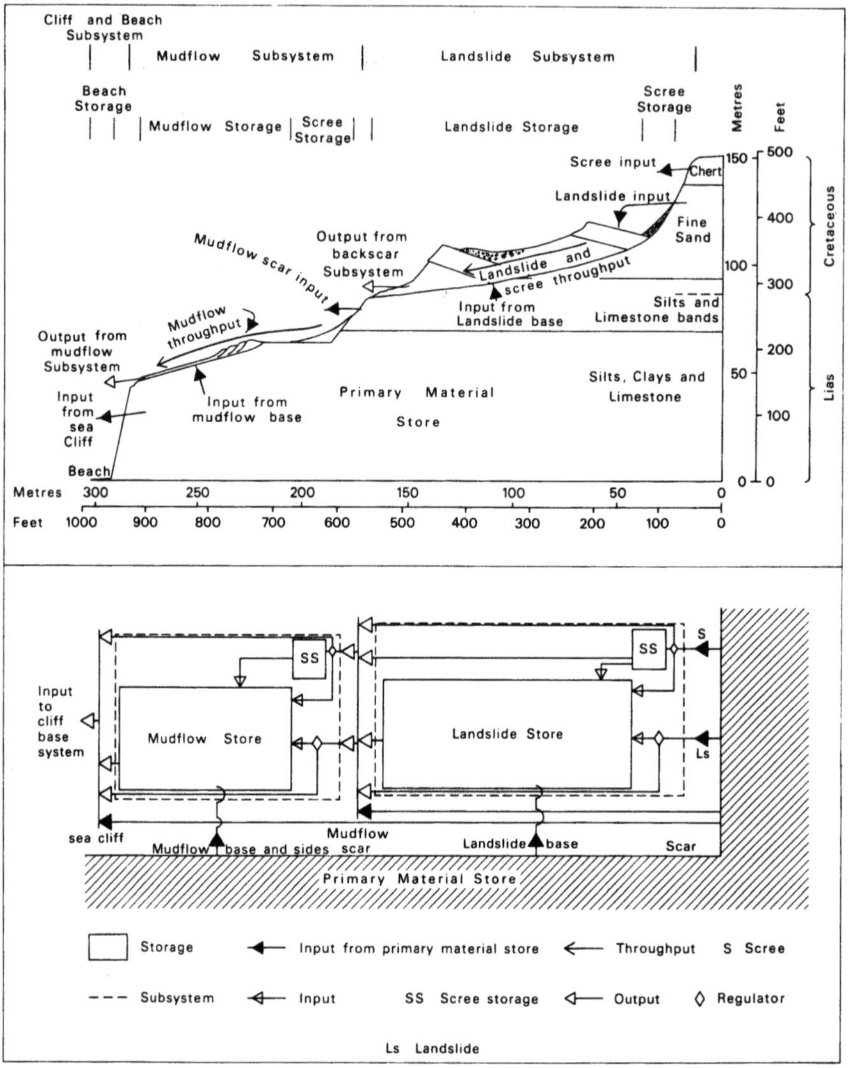

1.10 Schematic diagram of a cascaded landslide system in Dorset (after Brunsden 1973).

The systems are characterized at any particular time by the state variables of the system. In a soil system, the state could be described in terms of soil thickness, grain size and chemical composition. These characteristics are the result of the history of the system in recent or distant times. The states of some systems depend only on the immediate past and are said to have a very short memory. For example, dissolved solutes in a tidal pool would be completely independent of the dissolved solutes in the pool in the previous ebb. Such a system has no memory. Insofar as soil moisture on a particular day may still be quite high as a result of an earlier rainfall, the soil moisture on the same day after a second rainfall would still be, in part at least, related to the rainfall of the first storm. These situations are usually termed *antecedent* conditions. On a much longer time scale, one may anticipate that the slope form in one year would supply a strong 'component' or 'influence' on the slope-form a year later. The landscapes of resistant rocks seem to have quite long 'memories' and it is on this basis that geomorphologists undertake studies of past climates, deposits and erosion surfaces. This has important implications for theoretical work. Notice that here we are implying that if an input were a map of elevations at time t and the output a map of elevation at $t + k$, where k is the time elapsed between the two maps, then the map at t will in part determine the map at $t + k$. In a spatial sense outputs from one system may become the input for another and the systems are said to be cascaded (fig. 1.10).

Table 1.1 Status of variables during designated time spans (from Schumm and Lichty 1965)

		Status of variables during designated time spans	
Drainage basin variables	*Cyclic*	*Graded*	*Steady*
1 Time	Independent	Not relevant	Not relevant
2 Initial relief	Independent	*Not relevant*	*Not relevant*
3 Geology (lithology, structure)	Independent	Independent	Independent
4 Climate	*Independent*	Independent	Independent
5 Vegetation (type and density)	Dependent	Independent	Independent
6 Relief or volume of system above base level	Dependent	Independent	Independent
7 Hydrology (runoff and sediment yield per unit area within a system)	Dependent	*Independent*	Independent
8 Drainage network morphology	Dependent	Dependent	Independent
9 Hillslope morphology	Dependent	Dependent	*Independent*
10 Hydrology (discharge of water and sediment from system)	Dependent	Dependent	Dependent
River variables	*Geologic*	*Modern*	*Present*
1 Time	Independent	*Not relevant*	*Not relevant*
2 Geology (lithology and structure)	Independent	Independent	Independent
3 Climate	*Independent*	Independent	Independent
4 Vegetation (type and density)	Dependent	Independent	Independent
5 Relief	Dependent	Independent	Independent
6 Paleohydrology (long-term discharge of water and sediment)	Dependent	Independent	Independent
7 Valley dimension (width, depth and slope)	*Dependent*	Independent	Independent
8 Mean discharge of water and sediment	Indeterminate	*Independent*	Independent
9 Channel morphology (width, depth, slope, shape, and pattern)	Indeterminate	*Dependent*	Independent
10 Observed discharge of water and sediment	Indeterminate	Indeterminate	Dependent
11 Observed flow characteristics (depth, velocity, turbulence, etc)	Indeterminate	Indeterminate	Dependent

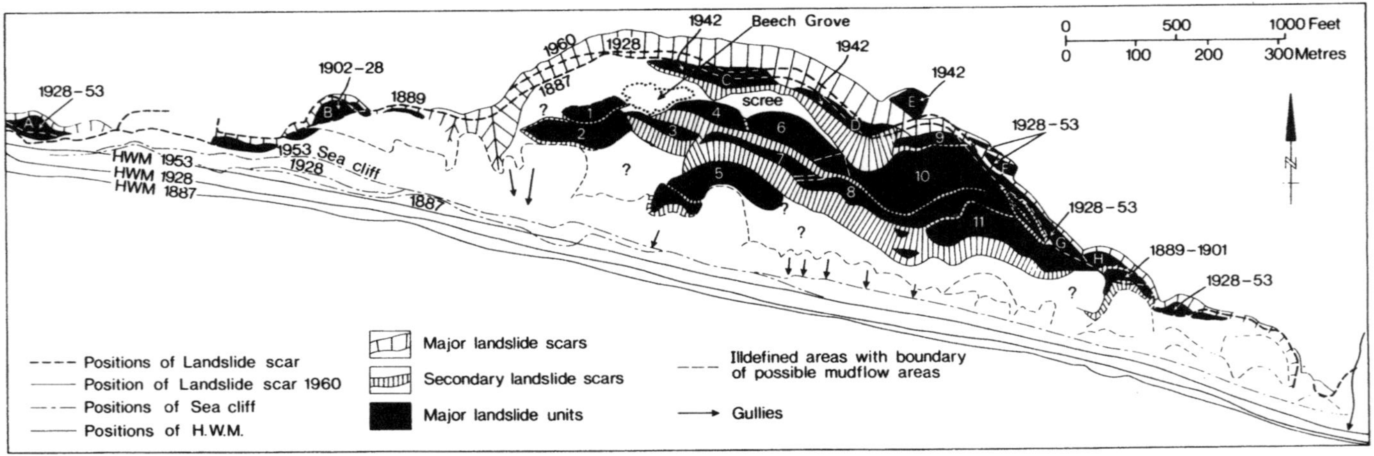

1.11 The spatial variability of mass movement between 1887–1960), Stonebarrow, Dorset (after Brunsden and Jones 1975).

Another vital temporal property of systems is that, if their behaviour is observed over a period of time, the output may be very steady; that is the fluctuations in the output are such that over the observed period the mean and variance of the variables which describe the output remain constant or statistically stable. The sediment yield of a river observed from month to month could remain statistically stable over a period of say ten years. If, however, for a Mediterranean country the sediment yield was observed for the month of July, the sediment yield might change rapidly from day to day. The stability or otherwise of a systems output, in other words, can only be specified in terms of some particular time scale. This important point was only generally recognized in geomorphology after publication of Schumm and Lichty's paper (1965). These authors pointed out, at the same time, that the controls of a system would also change in relative importance through time (table 1.1).

It has also been recognized that there is a very large gap in our knowledge between rates of operation of processes derived from studies of perhaps one to twenty-five years duration and gross figures of process operation on a geological time scale. There is an obvious temptation to use short-term figures to obtain long term effects but unless we know the variability of the processes involved such calculations may be meaningless. A future task in geomorphology is to determine the length of observation required to incorporate fluctuations in status variables for various processes. We have even less idea of the spatial variability of processes and until we do we will be unable to place reliance on our gross figures or adequately to predict future events (fig. 1.11).

Superimposed on a static or statistically stable systems output at one time scale there may, on a longer term, be an overall trend. The time scales referred to here may be assumed, for convenience, to be those of Schumm and Lichty (1965) who proposed the terms cyclic, graded and steady time for periods of the order of 10^6, 10^2 and 10^{-2} years respectively (table 1.1). The short term condition of a river may be a relatively stable condition upon which is superimposed, over the long term (10^6y), reduction of its landscape and lowering of its channel. In some explanations long-term change may be envisaged as a gradual and smooth process. This is not too realistic a picture since, more often, relatively sudden changes have occurred in the input of this system by way of changes in the relative level of land and sea. The Davisian model of geomorphology expresses the view that rivers adjust regularly and throughout their courses to such 'shocks' but the time taken for this adjustment will vary from process to process and place to place.

On a shorter time scale, say graded time, the 'shock' to a river channel system could be, for example, a dam burst. More naturally, an Icelandic *jökullhlaup*, in which changes in the internal drainage pattern of a glacier leads to large-scale drainage of a post-glacial lake, would provide such a sudden and dramatic shock to the system. Thereafter, the system operates through time to restore statistical stability. In other systems it may be some internal operating mechanism of the system which fails rather than a change in the input. This is most frequent where strength thresholds exist internally.

Figure 1.12 shows the rate of movement of an Antrim landslide (Prior and Stevens 1972); when the limiting pore water pressure internal to the material is exceeded the whole system changes to produce a completely new system of relationships which is sub-critical. In landslide systems the attribute of most interest might be the shape of the cliff profile; the effect of the landslide is to change the whole of this profile to a rather irregular one. The natural tendency of the system is to return this to a straight slope.

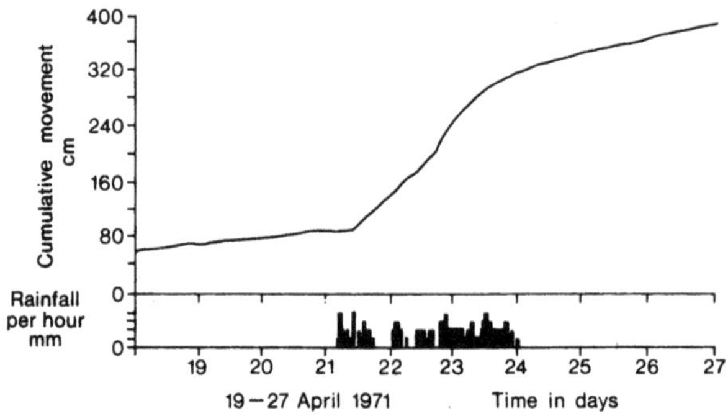

1.12 Dynamic response of an Antrim mudslide to precipitation after a threshold has been exceeded (after Prior and Stephens 1972).

The time period over which such adjustments take place is known as the *relaxation time*, the time taken for a system to achieve a new equilibrium following a change in the input or in the internal operation of the system. If a valley wall supplies material to the slopes below, talus will be created provided that the material is not rapidly removed. Suppose a well-sorted, rectilinear talus slope has developed on a stagnant ice mass (fig. 1.13a), and conditions of supply with no basal removal exist. Rapid melting of the ice mass may then cause changes in the profile, exposing bedrock or till on the slope. Eventually if the supply or input at the head of the scree continues the slope will gradually readjust to a new equilibrium condition. The time which this takes is the relaxation time of the system. The state of the system could be described by the x and y co-ordinates of the profile. The successive states (profiles) which the slope adopts after the shock is the *relaxation path*. If a series of talus slopes, subject to melt and collapse at progressively younger periods, were to be observed (fig. 1.13b) one might expect the same sequence to occur. That is, we substitute a spatial relaxation path for a temporal one (Thornes 1971).

The relaxation time for a system varies from one system to another. Some idea of the length of the relaxation time for a geomorphological system which is relatively rapid in its adjustment to new conditions is seen in the process of infilling of a river channel after a flood (figs. 1.14a,b), but more generally the time required is fairly long if complete adjustment is to be achieved. Modern work (Brunsden and Kesel 1973) has suggested that the process may be one of

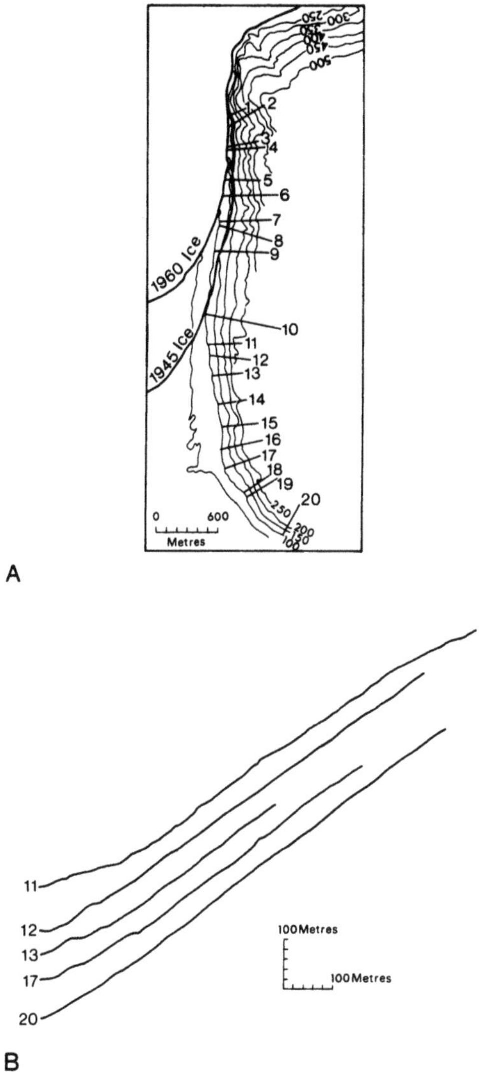

1.13 (a) Position of slope profiles relative to a retreating ice front in Iceland showing changing basal conditions. Note: some contours in the lower part of the map are omitted for clarity. The correct profile height is shown by the profiles in fig. 1.13b. (b) The successive states and relaxation path assumed by the talus slopes through time from retreat of the basal ice (Thornes 1971). Changes are most noticeable in the lower parts of the profiles.

initially rapid change followed by long periods of decreasing activity (fig. 1.14c).

In general, the relaxation time will be longer if there is more resistance to change in the system. Slopes on unconsolidated materials, such as sands and

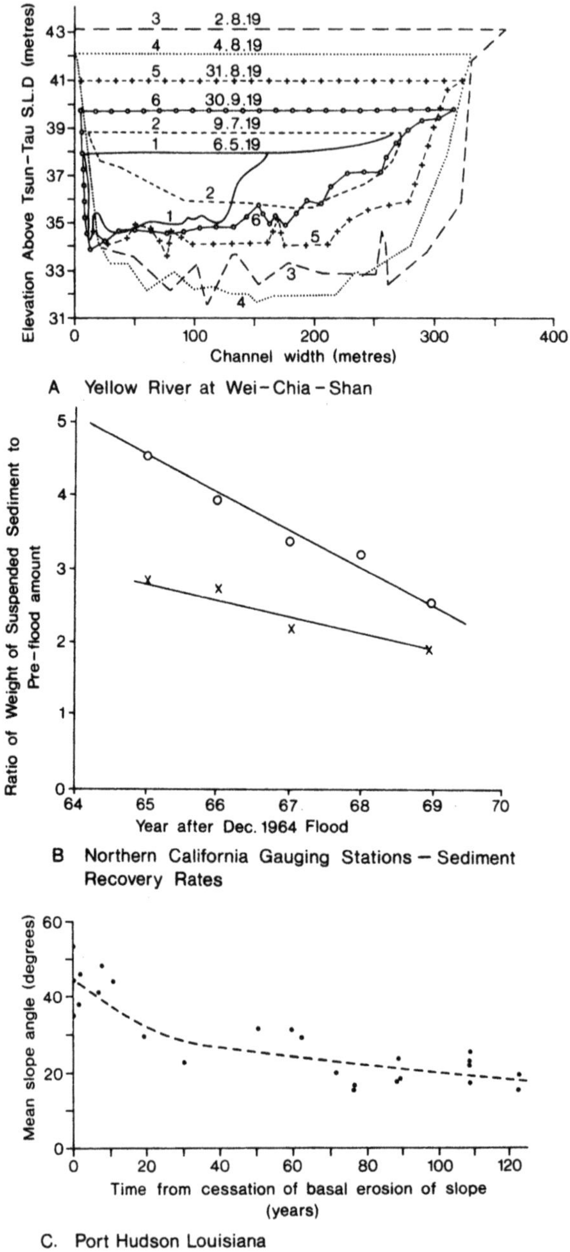

1.14 Systems adjustments after 'shock' indicating the magnitudes of the relaxation times. (a) Successive cut and fill of the Yellow river channel at Wei-Chia-Shan during a flood (Freeman 1922). (b) Sediment recovery rates in a channel following a flood (Anderson 1971) for two stations in Northern California. (c) Response characteristics of slope angle following cessation of erosion at the base of the slope (Brunsden and Kesel 1973).

gravels and marls, adjust rapidly to changes in the channel gradient; whereas on more resistant rocks, limestones, cemented breccias and sandstones, adjustment is much slower. As a result there is a strong correlation between channel and maximum valley side slope angle for the less resistant lithologies, and a weak correlation on the more resistant. Wild fluctuations in the input of the system also seem to even out more quickly and more effectively in more complex systems. This is because the system acts as a filter on the input, dampening down the input and imposing an autoregressive or moving average effect on it (see chapter 3). Where there already exists a strong memory effect in the input, the relaxation time of the system is again likely to be appreciably longer. The very short memory models are called probablistic or stochastic models, for they have a strong random component. Long memory systems are generally regarded as deterministic. In chapter 3 the different ways of measuring and characterizing geomorphological systems memories will be discussed

Evolution in time

The third major area of investigation which has time as its central theme is that of the evolution of landforms in which we seek to explain a phenomenon by way of the events which preceded it. There seem to be three main reasons why this should have been and continues to be a vital theme in geomorphology. First, most environments have long memories; they carry in them vestiges of past erosional events for at least several millions of years. Landscapes are not 'clean' pure products of contemporary processes but have in them a background of residual effects of earlier periods. These residues act to obscure the effects of contemporary processes, and act as constraints upon these processes. It is true to say, however, that the residues exist most strongly in areas of little contemporary activity since erosion is obliterative in its action. Conversely, in active areas it is only because deposition has taken place that some clues to the past history of an area are preserved.

The second reason is that the history of the landscape seemed to have in it the only possibilities for the testing of ideas about contemporary processes in the long term. Unfortunately, this argument has an element of circularity which went unrecognized for a long time. Geomorphologists and geologists sometimes tended to base their understanding of process on the study of form and then to seek elsewhere to confirm their beliefs by reference back to the form itself. The third and perhaps major reason was the influence of the principle cyclical model of landscape evolution of W. M. Davis (1909). His model was, above all, a time-decay model in which the passage of time could be observed in the form of the land and in which the passage of time in particular dictated the gradual reduction of relief. The model has been discussed at great length elsewhere (for a recent review see Chorley, Dunn and Beckinsale 1973) and we shall not dwell on it here save to say that it is a special case of a more general model in which uplift and denudation compete to produce and remove mass. Most criticisms of the model arise from (i) the rate at which mass is produced, (ii) the way in which mass is removed, and

(iii) the relaxation path along which a new most probable situation is reached.

However, although Davis was largely responsible for the type of denudation chronology practised in Britain and eastern North America in the first part of this century, the roots of the evolutionary school go back well before this, as Chorley, Dunn and Beckinsale (1964) have shown, and advanced on a broader front. It was implicit in almost all geology concerned with the great national surveys; it was closely allied with the development of glacial geomorphology and was, and continues to be, one of the roots of climatic geomorphology.

Three examples will suffice to illustrate the wide application of evolutionary ideas in geomorphology. In the first from classical denudation chronology, Wooldridge and Linton (1955) endeavoured to interpret the development of the landscape through time from morphological evidence. It was Wooldridge's belief, in particular, that geomorphological studies of erosion surface and drainage pattern sequences could be used to elucidate the

Table 1.2 Classic model of the denudation chronology of south-east England (Wooldridge and Linton 1955)

Time	*Events*	*Relicts*
Cretaceous Cenomanian and Senonian	Complex sequence of subsidence and uplift during which the Cretaceous rocks were deposited, uplifted and eroded. Production of ancient peneplain	Exhumed, inclined surface beneath remaining Eocene deposits, occasionally on stripped plane
Tertiary Early Tertiary	Submergence, trimming of peneplain Deposition of Tertiary deposits	
Mid-Tertiary	Uplift and folding in association with Alpine movements. Sub-Eocene surface tilted. Initiation of new erosion cycle	
Mid-Late Tertiary	Production of peneplained land surface	Mio-Pliocene peneplain — summit bevel
Pleistocene Calabrian	Submergence of part of Late Tertiary surface. Production of a marine bench	'600' foot platform Calabrian surface 550-700 ft
To present	Dissection, stage-by-stage, of the sea floor and Mio-Pliocene surface with successive marine incursions and retreats, at successively lower levels, together with low level raised beaches and buried channels	Shorelines and river terraces below 700 ft Buried channels

latest stages in geological time, and to date the events of the Tertiary and Quaternary. The general model set up by Wooldridge is shown in table 1.2. The techniques used are described in greater length in chapter 5. The second case relates to the development of slopes in the region of Exmoor, south-west England (fig. 1.15a). Carson and Petley (1970) argue that there are three principal modes in the distribution of slope angles. Each of these is believed to represent the results of the degree of mechanical breakdown in the debris

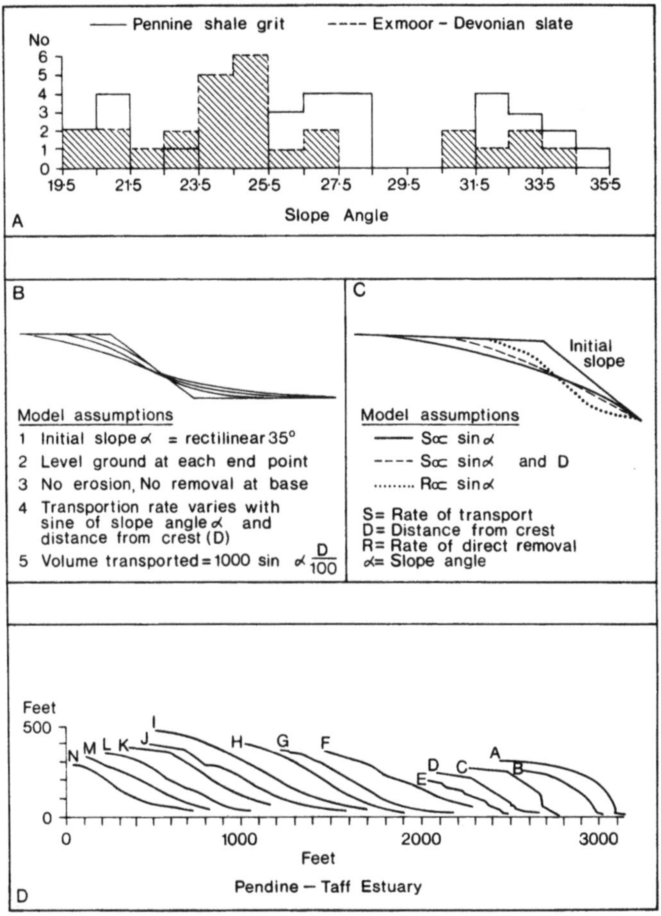

1.15 (a) Histograms of slope angles on Pennine grits and Devonian slates (Carson and Petley 1970). (b) Deductive model of slope evolution developed by Young (1963) to illustrate the development of accumulation features as a consequence of downslope soil transport. The model assumptions are shown in the diagram. (c) Deductive models of slope evolution developed by Young (1963) for three sets of model assumptions. The basic conditions include rectilinear initial slope at 35°, level ground above, no erosion and impeded removal at base of slope. Each model has variable transport and removal rates as shown in the diagram. (d) Slope profiles showing suggested change of form following the retreat of the sea from the base of a slope in South Wales (after Savigear 1952).

cover. They suggest that the evolution of slopes takes place in three phases by rapid change between these modal values after a threshold has been passed, and that slopes stay for longer periods of time at the next modal values. In the third case, one of Young's deductive models of slope evolution (Young 1963) (fig. 1.15b), the conditions whereby a slope must develop through time are set up, and the results of these conditions are subsequently observed by simulation of the process.

Here are three very different ways of incorporating passage through time into the understanding of the present landscape. In the first, an empirical study, extensive observations of erosion surfaces, uncomformable geological relationships such as the sub-Eocene surface and the relationship of drainage pattern to structure, *in the existing landscape,* are used to induce a schematic history of the landscape. Corroboration of this schematic history is then sought in physically similar situations elsewhere. In the second case, the spatial variations which occur in straight slope angles on Exmoor are explained in terms of a relationship with the mechanical characteristics of the weathered debris under a set of process assumptions. The hypothesized mechanism of the breakdown of debris through time then leads to a scheme of slope development through time. This argument employs some aspects of a space-time transformation (see below). The third and final model is a deductive and iterative model. The author deduces what he considers to be the principle controlling variables in the system and watches as the system evolves under these constraints and in isolation from nature. His deductions are based on a knowledge of contemporary process, but the model attempts to replicate what happens over time. The first case is based on what is believed to have happened in the past; the last on what is happening at present and the second on a mixture of the two. They all have in common the changes of form over time. They also demonstrate the use of evolution as a history or sequence and not a process in the Darwinian sense.

The models are different in that they rely to a greater or lesser degree on information about events occurring in the past. In the first, the reliance on the historical record is absolute both for the derivation of the initial model and for its confirmation. We might expect, therefore, that the principal difficulties lie in the piecemeal nature of the information. The procedure adopts two assumptions which yield further problems. One is that particular forms relate to particular processes or assemblages of processes; the second that the nature and rates of operation of processes in the past were much as we observe them today. Both these assumptions have been attacked on the grounds that the particular end form can be reached along many paths, and that we are not always able to choose between them. A flat plain might be produced by marine abrasion, subaerial denudation, fluvial aggradation, exhumation and perhaps in other ways. Hopefully, however, we might expect other kinds of supporting evidence, such as marine or fluvial gravels, and degraded cliff lines to help us decide.

In terms of slope form the criticism is rather more severe. The second assumption relies on the fact that every conceivable geomorphological environment and process in all their possible combinations can be found now on the earth's surface. The main criticisms are:

(a) that all environments on the earth have now been substantially affected by man and this is especially so where the relationship between process and form is being most actively investigated, and
(b) that although analogues do exist our knowledge of them is mostly very superficial. This is especially true if temporal and spatial variability is considered.

The main problem of the second case study is the difficulty of verification. This problem arises equally in the third case study and many others like it. Culling (1965) discusses this problem at length with respect to his proposed theory of soil creep. Following Popper (1965) he points out that a theory is scientific in so far as it is potentially falsifiable, and goes on to discuss the methods of seeking corroborative evidence for the theory. Geomorphological evidence suffers from the facts that:

(i) only in very favourable circumstances is it possible to determine the initial and boundary conditions;
(ii) that even when initial conditions are available they almost invariably have to be plane surfaces because any less regular surface can never be known with sufficient accuracy;
(iii) even in the most favoured and well-known localities our knowledge of past conditions falls far short of the accuracy needed for significant assessment; and
(iv) any theoretical problem suitable for a comparison with those of the real landscape will be sufficiently complex to rule out any solution by analytical methods.

The difficulty of matching necessarily simplified but analytically tractable models with the evolution of complex landforms, from very imperfect knowledge, is most acute when dealing with the evolution of form. In the last case example above, the question is not whether the model operates correctly or even whether the results are 'right' given the starting conditions. Rather it is whether the starting conditions are correct and if there are adequate means of comparing the results with those of the real world.

One approach to this general group of problems is to argue that, given sufficient information, one should simply forecast the landscape on the basis of knowledge of existing processes. If the model can be calibrated against the outcomes of present day processes then it can simply be extrapolated to the longer term. In short, assume that the model is self-testing and do not appeal to the past for verification. The problems of achieving this rest on our ability to discover the temporal and spatial variability of the processes. Another approach is to seek for verification by using an ergodic transformation in which, because the relaxation time of many geomorphological systems is long in relation to the human time scale, we attempt to explain distribution in time by recourse to distribution in space.

Imagine a person starting out into a complex network system from which there is no exit. If he now moves across the links of the network, tossing a coin at every junction, we could estimate the probability of the man being in, say, link E after time K. That is we could obtain as a percentage the probability that he would be in link E. We could also express the probability that he has spent so many minutes in link E.

Another interpretation of this could be that if a large number of people started out from the entrance to the network, after they had wandered round for a sufficient time, then the probability of a person being in link E could be expressed by the ratio of the number of people in E to the total number of people in the network. In other words, by looking from an aeroplane at the spatial distribution of people in the network, we could also say how much time an individual would spend in each link if he were to have been circulating around the network for a long time. This is the formal idea of the ergodic hypothesis.

The first interpretation involves one person and asks what was the relative amount of *time* spent in each link; the second says what is the relative *number* of people in that link in an instant in time after equilibrium is reached. Where these are the same when expressed as probability, the system is said to be ergodic.

The same idea extended to landforms would be formally represented by saying, for example, given that 25% of all river channels exhibit the 'youthful' characteristics of Davis' ideal river, 50% the 'mature' characteristics and another 25% 'old-age' characteristics, then we might expect a single river, in the course of its 'life' to spend 25% of its *time* having youthful characteristics and so on.

A large danger of distortion of this idea comes quickly to mind. Suppose we have some particular idea about slope evolution; perhaps that slope declines rather than retreats parallel to itself. If we observe a set of slopes and put them in order of slope angle such that they match the hypothesis, we have neither tested the hypothesis nor made an ergodic transformation. But if we have a truly ergodic system, and we observe that system in a particular state, and this state is rare among all the spatially observable states, we could conclude that the particular state observed would be short-lived in the history of the particular system. Assume that talus slopes are ergodic systems. Then, for the ergodic condition to be satisfied, a peculiar talus slope form in nature (i.e. one that is relatively rare), would have to be relatively short-lived.

Perhaps because we do not know if our systems are ergodic in general the notion is used in the substitution of a spatial series in which the data can be arranged in a time sequence for true evolution through time (Savigear 1952, Strahler 1950, Carter and Chorley 1961, Schumm 1956, Simonett and Rogers 1970, Welch 1970, Chorley and Kennedy 1971, Brunsden and Kesel 1973). In Savigear's work, for example, a series of slope profiles were arranged in order of age as judged by the time the sea left the foot of the slope and basal activity ceased (fig. 1.15d). The series so developed was considered as analogous to a temporal series and the differences in the profiles taken to represent differences which could occur through time. This example is slightly confusing perhaps, because the arrangement of the slopes in a time-order was a function of their spatial position. This need not be the case. 'Space-time analogue' might be a better term than ergodic transformation. In a later chapter (p. 164) we shall want to use the ergodic notion in its true context in connection with Markov Chain modelling.

A second approach to the verification of evolutionary models is to simulate the real situation either mechanically or electronically, with the hope of speeding up time so that the hypothesized succession of events may be observed. Such models are also used in an attempt to understand the internal

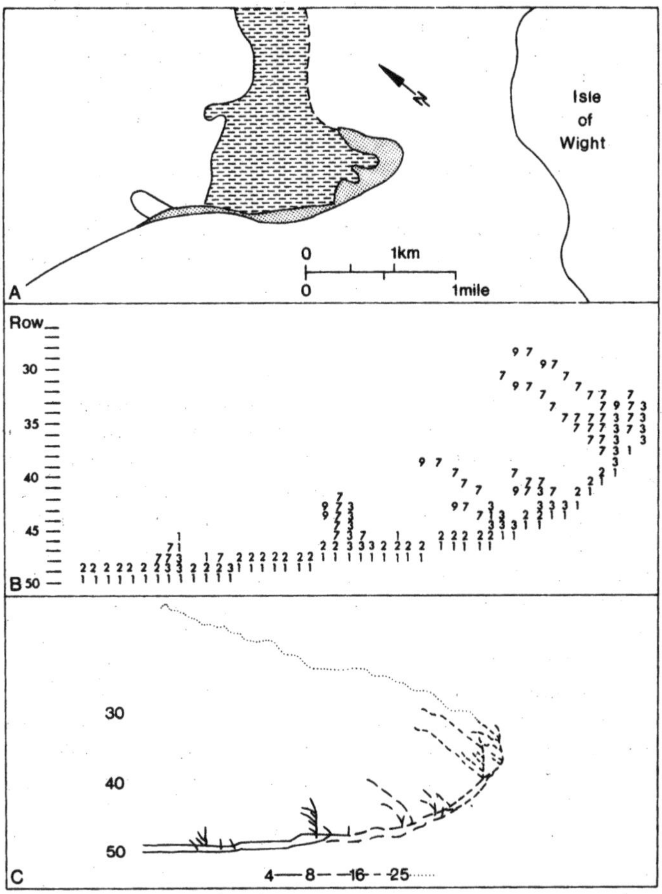

1.16 The development of Hurst Castle Spit simulated by computer. (a) Map of Hurst Castle Spit; stipple = spit, broken lines = marsh. (b) Computer output. The numbers represent the particular types of process events and sequences required to produce the geometry and morphology of the spit. (c) Best-fit form with successive iterations, shown by different symbols (after King and McCullagh 1971).

workings of the systems. An example is provided by the simulation of spit development attributable to King and McCullagh (1971). Here (fig. 1.16) the development of Hurst Castle Spit was simulated by a computer model based on assumptions about the offshore topography, the principal wind direction, the rate of supply and several other conditions. This model involved deterministic simulation, that is the evolution of the spit continued according to fixed and immutable rules. Every 'run' of the model was replicated by inputting the same conditions and operating subject to the same rules. In chapter 7 we shall show that in some simulation models the processes are regarded as probabilistic. In both types of model, the system is speeded up by use of repeated operations at a rate much faster than in nature. An example of physical simulation which speeds up the rate of operation of the process is the use of estuary models to determine factors controlling siltation. In such models tidal cycles, both diurnal and seasonal, are speeded up so that evolution of bars or spits can be traced.

A somewhat similar procedure is to obtain field situations in which operations naturally take place at a much faster rate than usual. These are then used as analogues for the slower rates of operation on a larger scale and on resistant lithologies. Schumm's (1954) classical study on erosion in badland areas demonstrates this technique.

Other time-based issues

This chapter has concentrated largely on magnitude, frequency and rate, geomorphological systems and time, and has outlined some of the basic notions and problems associated with evolutionary studies. Though these are among the most important issues in which time is central they are not the only ones. Little has been said about forecasting, for although it is closely linked with geomorphological systems analysis, the retrospective rather than prospective view in geomorphology has meant that little work has been done in this area. Yet much is required; forecasting sediment yield is strongly tied to the economic life of reservoirs; forecasting soil loss has fundamental importance for agriculture. Moreover forecasting has not only purely economic significance; forecasting ability suggests ability to control; controlled experiments by their very nature yield more information both about the system itself and about the controlling agent. A recent review of these problems, applied to drainage basins, is that of Gregory and Walling (1973).

Another issue of which relatively little has been said and done is in the idea of storage. This too, albeit somewhat artificially, can be tied in with geomorphic systems but it is worthy of discussion in its own right. In relation to time, storage is significant as a form of inertia in the system, tending to stabilize output, releasing at times of scarcity, absorbing in times of excess. Storage and rate are closely interrelated and hence storage involves time. As an example, think of soil as being the stored products of weathered bedrock. If the rate of removal by running water exceeds the rate of production at the weathered interface and that supplied from further upslope, then the first change in state is towards depletion of the store and the soil cover becomes thinner. The store acts as a buffer and is conditioned by rates of operation. In a later chapter, we will show how, in stochastic models, storage may be represented essentially by time-based functions.

Finally, we have to remember that processes are simply combinations of circumstances which change the state of our system over time. There is no such thing as a time-independent process, though there can be processes whose effects over time are negligible. In this case either we are not aware of them because their effects are not observable, or they are unimportant because they produce no change. We can only conclude that differences of view about the significance of time in geomorphology are, in substance, differences of degree rather than of kind.

2 The measurement of time

Most geomorphological studies involve measurement of the passage of time usually, though not necessarily, with respect to the position of events (epochs) in the calendar scale. Many geochronological studies are inseparable from studies of past environments, climatic fluctuations or determinations of the rate of geomorphological processes. The results of such studies have important implications for applied geomorphology as well as for the more academic aims of establishing the sequence, frequency and magnitude of geomorphological events, of establishing erosion chronologies, the effects of climatic or sea-level change or the mechanisms of drainage pattern evolution.

Two general methods of measurement are employed: *relative* and *absolute* position in time. The first includes those techniques, from many disciplines, which attempt to place events in some order without accurate time calibration. It seeks to establish the order of events but not the absolute duration either of the event or of the periods between them. The second method involves the determination of the number of calendar years since an event occurred and includes more precise methods in which events are related to a scale whose units are of equal incremental size. Relative techniques may be thought of as ordinal scale measurement with absolute dating as the ratio scale. The latter usually forms a more precise framework and 'spacing' of events for the former.

Relative position

The most common relative scale-dating techniques used in geomorphology are morphological, where spatial measures are used to determine relative temporal occurrence in the landscape; and stratigraphical, which considers the attitudes, discontinuities and physical properties of geomorphological deposits and includes semi-precise techniques such as the study of preserved flora, fauna and pollen, fossil soils and volcanic ash sequences and artefacts.

Morphological techniques

The crudest and in many cases most unreliable techniques for determining relative position in time have been those involving height, or relative height.

Absolute or relative height has been used to date large-scale erosion surfaces, sequential river terraces, wave-cut platforms, glacial cirques, moraines, cave systems and many other results of geomorphological activity. A few examples will illustrate the general proposition and highlight some of the pitfalls.

Erosion surfaces and altitude. In river terrace chronology the most important criterion for correlating terrace remnants or erosional benches is based on the assumption that during a period of relative stability a continuous surface was cut or built along the entire length of a valley. Later incision by the main stream and dissection by its tributaries reduces this surface to a few terrace remnants. Repetition of these events leads to a complex sequence of surfaces on the valley sides the height of which is then used to develop a terrace chronology. Two other criteria have generally been used: (i) the relative position of a terrace remnant in such a sequence (e.g. 'middle terrace') and elevation above the stream bed. A good example of the use of all three principles is found in Vita-Finzi's (1963) study of aggradation in the circum-Mediterranean valleys. Two valley fills are recognized in terms of relative elevation, the older fill being generally higher and more dissected. The younger fill, representing a period of renewed aggradation, is a virtually

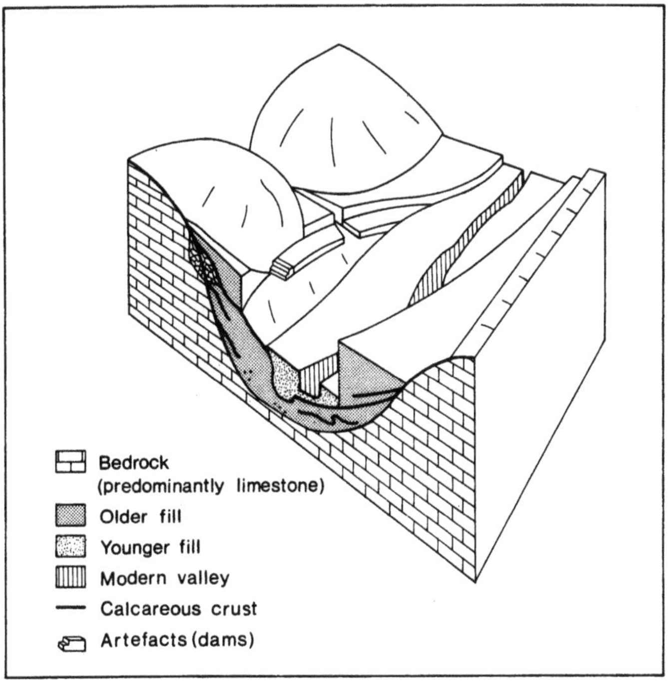

2.1 Schematic block diagram showing the final stage of the geological history of a Tripolitanian wadi (after Vita-Finzi 1963).

continuous surface in individual basins, such as the Wadi Deheb. The basis of the circum-Mediterranean correlation is essentially the height of the aggradational terraces and their relative position, supplemented as frequently as possible by reference to Roman artefactual material, notably earth dams (fig. 2.1).

In the cases examined by Vita-Finzi the strong aggradational character, continuity and lithological distinctiveness of the terrace deposits added strong confirmation to correlations based on elevation. In other areas, however, height, particularly when taken alone, is a dangerous criterion. Changes in run off and sediment supply in a stream system may lead to breaks in valley slope and therefore to abrupt steps in future terrace height. A corollary is that a smooth concave profile of regular mathematical form is not essential for the description of channel stability. Best-fit mathematical curves of channel form, a technique widely practised in denudation chronoloy to correlate terrace fragments (Jones 1924), do not necessarily yield a contemporaneous former flood plain. Continuity of a surface down a valley is the prime morphological criterion, rather than height, but it must also be remembered that if a surface was formed during a period when a stream was progressively downcutting then a single continuous surface will not be produced. The contemporaneous features will probably be isolated flat spurs, at *different* heights on the valley sides. Irregularities in deposition rates will only serve to emphasize this effect. In addition, by using height, relative position and height above the stream channel *alone*, we are unable to distinguish adequately the effects of isostatic readjustment, changes in the river regime and sea-level, and tectonic activity along the river course. These problems are well illustrated by the case of the Rhine in Europe whose terraces have been the focus of continued research for a long period of time. The main complications here rise from (i) the presence of a long section of intermittent tectonic activity between Basle and Frankfurt, and (ii) the effects of sinking of the North Sea coast together with fluctuations in Pleistocene sea level. Finally, there exists the problem that our knowledge of erosional surfaces of fluviatile origin is still the subject of much debate. The original Davisian schema suggests a sequence of surfaces evolving through time with successive surfaces, termed youthful, mature and old age, regarded as separate and isochronous. Other schemes, for example that of King (1953), envisage escarpment retreat away from a channel so that a single surface may be metachronous; the range of time involved increasing as the scarp retreats further from its point of origin. Dating in this case depends on a knowledge of the origin of the surface and is determined by the conceptual framework of the study! (fig. 2.2).

Sea levels and altitude. The problems of correlation on the basis of altitude are also admirably demonstrated by reference to former sea-levels. The general desire to date the shoreline platforms around the coast of Britain led to a classification of the phenomena on the basis of height and an assumption of age equivalence. Attempts have been made to correlate British terraces with those of the Mediterranean using height criteria (Green 1943). The

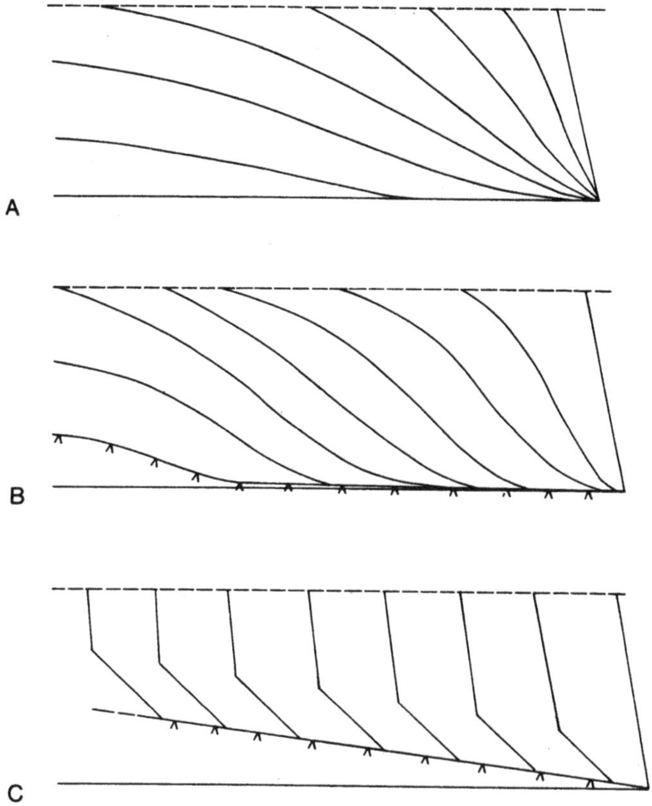

2.2 Examples of isochronous (firm lines) and metachronous (arrowed lines) surfaces produced by different schemes of landscape evolution. (a) Successive stages of decline. (b) Successive stages of slope replacement, but with lateral movement of slope base. (c) A model of parallel escarpment retreat yielding a metachronous pediment surface.

recent work in Britain, Arctic Canada and the pioneering work in Fennoscandia have shown the inadequacies of this assumption. In addition to the fact that isostatic rebound is not continuous and steady in time or space, the problems of differential ice retreat and sea-level change and local factors such as the construction and preservation of wave generated features or local tidal range add to the difficulties. Only with the development of absolute and better relative dating techniques have some of these problems come to light. The height of a surface must always be used with extreme caution (Frye and Leonard 1957).

Drainage patterns. Another tool which has been used in the absence of sediments to infer relative dates is drainage pattern development. Wooldridge and Linton (1955), in their study of structure, surface and drainage in southern Britain, used the degree of adjustment of drainage pattern to structure to infer that the drainage pattern and land surface of the Wealden 'Island' was of greater age than the surrounding area which, on the same evidence, was believed by them to have been levelled by marine erosion. In his study of till sheets in southern Essex, Clayton (1957) endeavoured to differentiate till sheets, in part at least, on the basis of drainage pattern development. That such a procedure seems at least feasible is illustrated by the comparative studies of drainage density made by Ruhe (1954) on till sheets of different ages in central USA. The effect on drainage density of the physical properties of the materials involved is, however, often neglected in such approaches.

Stratigraphical techniques

Wherever sediments exist, the somewhat dubious methods based on morphology alone give way to the rather more certain methods of stratigraphy in the strict sense. The relationship between erosion surfaces (especially unconformities) and correlative deposits is widely referenced in the geological literature, and here only three or four stratigraphic techniques of particular interest in geomorphology will be considered. The basic idea that the uppermost deposit is the most recent is usually satisfactory in recent sedimentation where (except in glacial activity) severe mechanical dislocation and overturning are relatively rare. The three most important stratigraphic tools are:

(a) organic remains, both floral and faunal; especially where these are linked to absolute dates;
(b) weathered products and relict deposits; and
(c) artefactual materials.

In many situations these techniques are used together and with other techniques.

Organic remains: palynology. Floral remains in sediments have yielded perhaps one of the most powerful stratigraphic indicators for the Quaternary and Tertiary periods. Much of this importance arises from the study of pollen grains (palynology), though macroscopic plant remains, particularly studied in the earlier part of the century, have proved highly important. The palynologist has as his main interest the reconstruction of the environments, especially the climatic conditions, in which the former plants thrived. An important by-product of this study is a most useful method of dating and correlation.

Flowering plants shed pollen and spores which are dispersed and trapped in sediments such as organic peats, lake muds and soils. The grains are very robust and will resist pressure, burning, transportation and the acids of circulating groundwater. This quality means that, for relict sediments, the pollen which was shed at the time of formation may later be collected and identified (Erdtman 1943). Extraction and counting of pollen grains from a

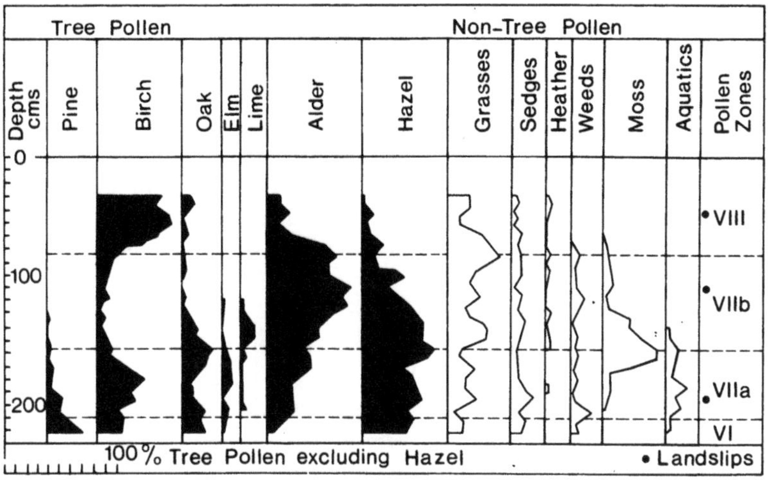

2.3 Simplified pollen diagram from landslide sites at Coombes Rocks, demonstrating the change in vegetation composition and its relation to post-glacial pollen zones (after Johnson 1965).

succession of horizons of the deposit yields a pollen spectrum which is generally characteristic of the regional environment during the time of formation of that deposit (fig. 2.3). In the pollen diagram arboreal pollen is expressed by species as a percentage frequency distribution of the total. Non-arboreal pollen and other spores are usually expressed in absolute numbers.

This very useful technique is not without its difficulties. The relationship of the contributing local plant association to the pollen being deposited at the same time is not one-to-one. Many species overproduce pollen while others may have a relatively sparse production. Differences in transportation mechanisms, and the possibility of reworked grains being transported over large distances or time intervals, suggest that the pollen frequency distributions may not be the same as that of the local source population (Davis and Goodlett 1960).

Some of these problems can be minimized by the comparison of contemporary floral assemblages with modern pollen rain; by allowing for varying rates of production and distances of transportation and by calibrating the number of each species in terms of pollen grains per unit volume of sediment. In a statistical sense it is important that several samples from a horizon should have less variance than samples from different populations (horizons). This implies that we should know something of the sources of variance. The laborious nature of the technique has, however, led to studies in which diagrams derived from just one core have been compared with diagrams from another single core from a location perhaps hundreds of miles away! The assumption that organic horizons separated by distance show less variance than horizons in the same deposit is scarcely valid until extensive checks have been made. Nevertheless, comparison of spectra, if used with care and where one of them is dated absolutely, say by radiometric techniques, provide a useful and relatively inexpensive method of dating, though one in which a good deal of training and skill is involved.

The sequence of post-glacial vegetational development is now extremely well known in Europe and in eastern North America. The picture for the glacial and interglacial periods is less well known but the technique is extensively used in geomorphological studies in many different situations. An example of the use of pollen stratigraphy in evaluating the course of late-glacial evolution is Brown's study (1962) of the development of late-glacial

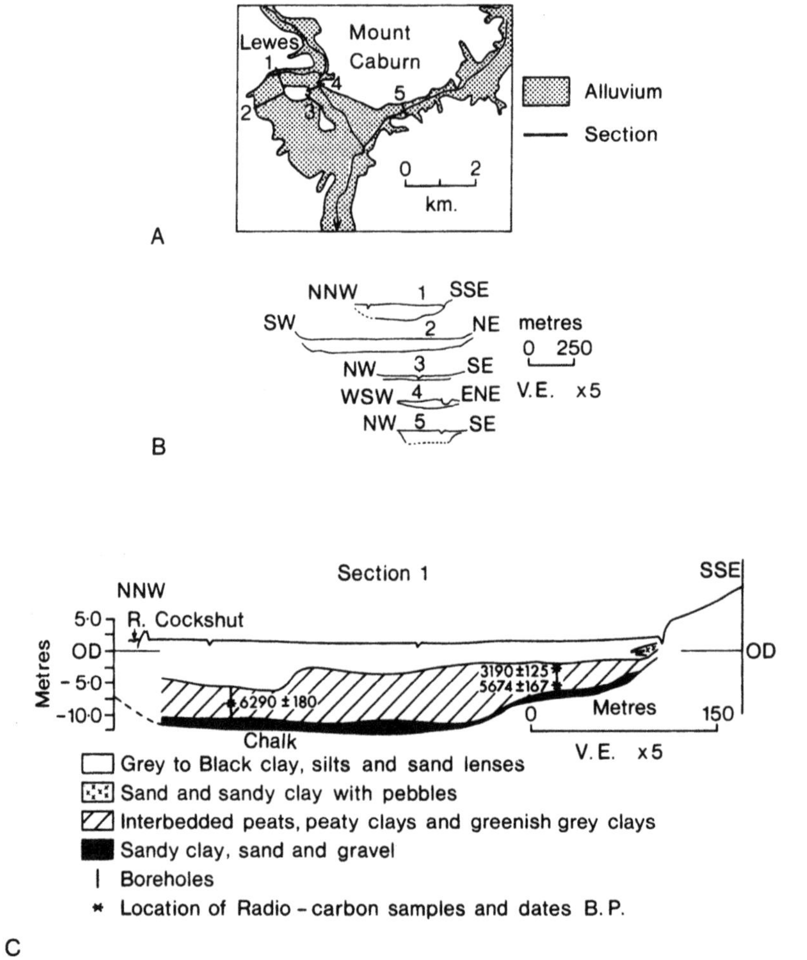

2.4 The alluvial fill of the Ouse valley, Sussex. (a) Location of alluvium and sections. (b) Simplifed sections of valley form. (c) Section showing distribution of deposit from which a pollen diagram and radio-carbon dates were obtained (after Jones 1970).

drainage in the lower Ottawa river–St Lawrence valley. Here the stages of occupance of major drainage channels in relation to uplift have been elucidated by using pollen diagrams with occasional radiometric dates. Johnson (1965) used pollen to date landslips of post-glacial age, and elsewhere it has been used to infer the ages of tills, raised beaches and alluvial fills. In the latter case, Jones (1970) (fig. 2.4) has been able to indicate the date and rate of aggradation of alluvial material in the Ouse valley in Sussex using pollen and radiocarbon dates.

Organic remains – faunal fossils. The use of faunal evidence for dating events in time is a similar technique in which a particular fossil or suite of fossils is identified as time-specific, so that discovery of this particular assemblage is taken as indicating sediments of that age. A wide range of species has been used for this purpose and the procedures followed are the techniques of stratigraphic palaeontology generally (Shaw 1964) in which interest focuses on the species present, index fossils, the spatial and temporal range of species and statistical methods of correlation. There are two basic points to be made about the use of faunal fossils. Because geomorphological interest very largely concentrates on the last few million years, a much more detailed subdivision of time is both possible and desirable than when compared with, say, the Jurassic or the Precambrian. A consequence of this is that more rapidly evolving species are required for precise dating. The second point is that terrestrial environments, which are both discontinuous and more highly specialized than marine environments, assume a much greater importance in studies of Quaternary and Recent deposits. In many respects this proves to be a real advantage, since without the oceanic buffer terrestrial fossils respond more rapidly and record less ambiguously many of the climatic parameters. In turn, the effects of climatic change have an increased importance as controls of landform genesis. This is not to say that marine species have not proved useful in dating. Temperature sensitive Foraminifera and marine molluscs have both been used for stratigraphic correlation; the former assumed particular importance with the development, since 1907, of radiometric dating together with deep coring techniques. The establishment of isotopic temperature curves from the fluctuation of O^{18}/O^{16} ratios is in some respects analogous to the pollen spectra mentioned above (fig. 2.5).

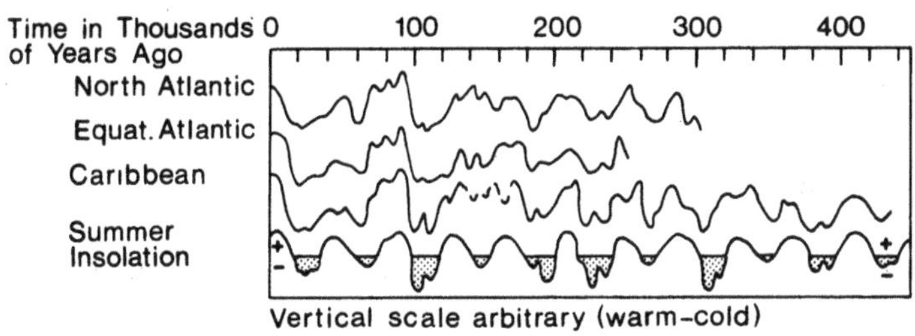

2.5 Averaged climatic curves shown by ocean cores in the north and equatorial Atlantic and the Caribbean (after Evans 1971).

Non-marine molluscs have been used, but interest here focuses more on the environmental implications since their evolution is relatively slow.

This raises the general issue that relative dating of events has been inferred from the environmental conditions supposed to have existed at the time of deposition. This problem is particularly acute in Pleistocene studies. So for example one often meets the assertion that because the fauna is 'cold' the environment was glacial and this is then used to date fairly precisely. This is sometimes satisfactory, certainly for the Late-Glacial and Post-Glacial pollen assemblages where climatic inferences may suggest relative position in time with a modest degree of precision. The practice is a dangerous one however, because few fossils and even fewer geomorphological indicators give *precise* information about environmental parameters and because we still have a very poor understanding of the role of climate in the evolution of most areas. The most extreme application of this type of procedure sometimes leads to a circular argument where, in one section, deposits are inferred to be of glacial age because of their character, while in another section, glacial origin of the sediments is inferred from their age, the latter being determined by reference to the first pit! This unfortunately is neither uncommon nor even easy to avoid especially where the inferences and assumptions are not specified in the literature.

As an example of the use of faunal evidence in geomorphology for dating purposes consider the use of non-marine molluscs in the British Quaternary. As Sparks (1953) points out, these animals appear far from ideal — their distribution is controlled by local conditions and evolutionary changes have been small or virtually non-existent. Nonetheless, Kerney (1963) has been able to establish the age of post-glacial deposits in south-east England on the basis of molluscan fauna, after tying in the stratigraphic record with radiometric dates (fig. 2.6). The species for each horizon are expressed in terms of absolute abundance per unit of sediment, comparable to pollen diagrams. One outcome of the work was the determination of the nature and main periods of geomorphological activity. The same techniques have been used to date coombe development, river terrace aggradation and sea-level change.

Weathered products and relict deposits. Sedimentary sequences and particularly weathering of sedimentary deposits have often been used to date geomorphological activity. Buried soils, wind blown loess and volcanic dust (tephra) are particulary important for Pleistocene and Recent events. Further back in time more conventional geological materials, correlative Flysch deposits, marine transgressive and regressive sediments and erosional incidences marked by unconformities, have been key tools of the denudation chronologists.

Palaeosols include relict soils (not subsequently buried) and buried soils. Those which are stratigraphically important have (a) widespread geographical distribution, (b) clearly defined characteristic features which serve as marker horizons, and (c) stratigraphic relations or organic content which permit fairly precise but independent age determinations in at least

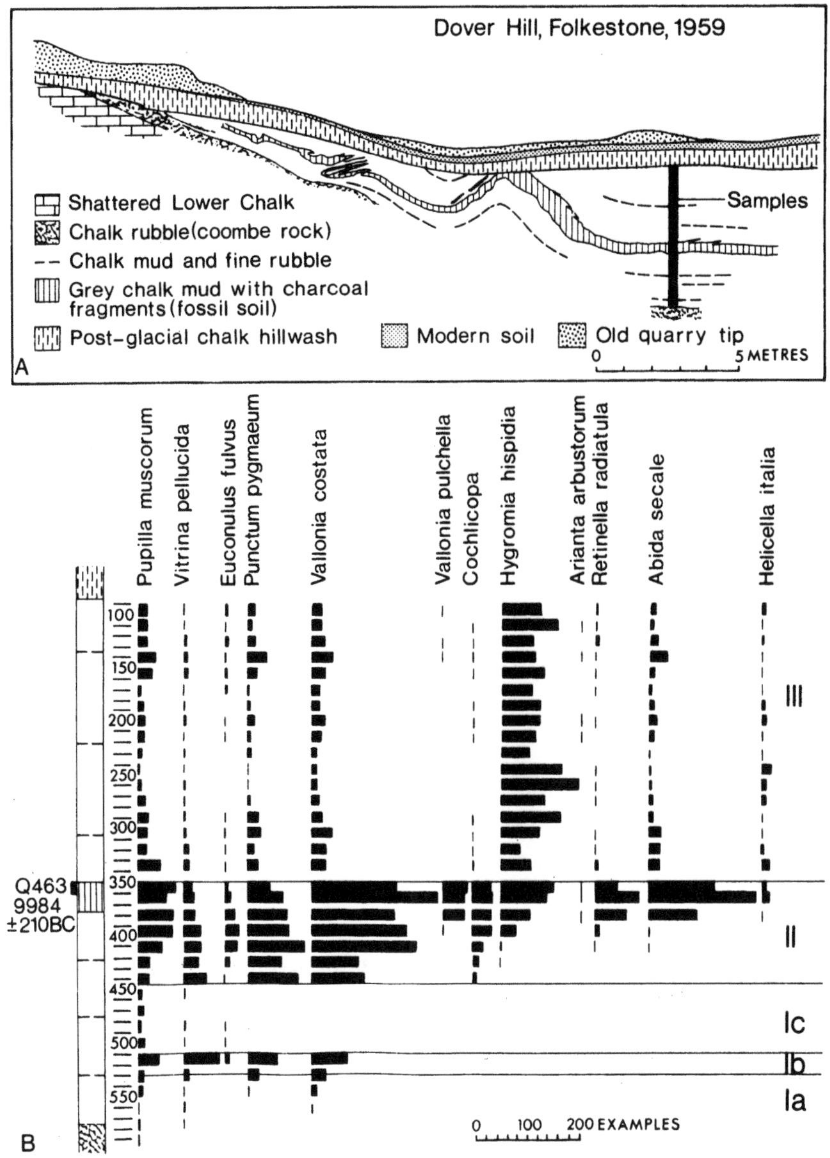

2.6 A molluscan fauna diagram with radio-carbon date for Dover Hill, Folkestone (after Kerney 1963). (a) Section and location of samples. (b) Diagram in composite histogram form.

some of its locations. From stratigraphic studies in the United States, palaeosols have been identified in a total of fourteen different horizons, ranging in age from Nebraskan to Recent. Palaeosols are *time-parallel*, that

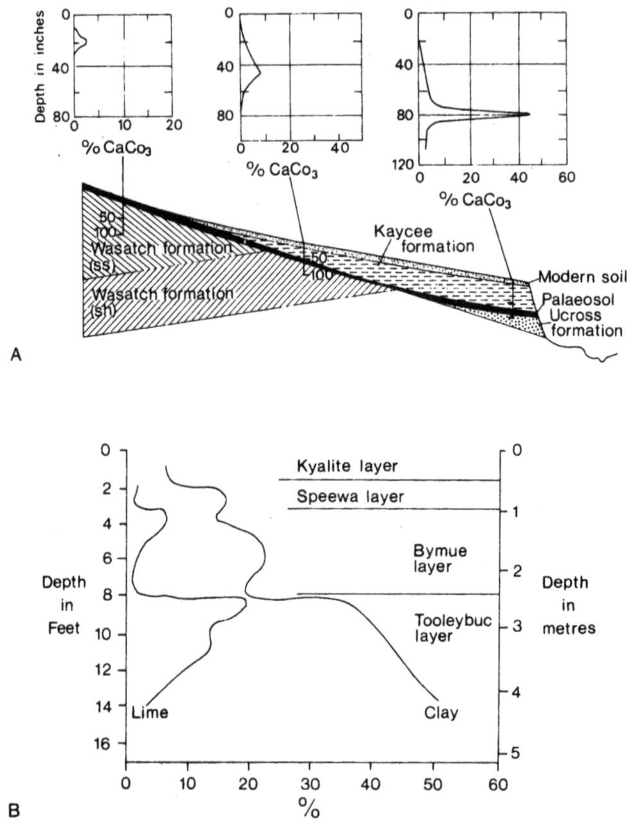

2.7 (a) Changes in $CaCo_3$ with depth for a buried soil in Wyoming (after Leopold *et al.* 1964). (b) Changes in soil properties with depth for buried soil layers (after Churchward 1961).

is to say, they are assumed (and often demonstrated) to be of equal age throughout their entire area. But it must *not* be assumed that a particular soil will remain laterally identical in physical properties; obviously the combined effects of varying parent materials and local climate, depth of burial and subsequent weathering, make this virtually impossible. Palaeosols have been, and continue to be one of the main criteria for differentiating between till sheets of different ages. One of the most important of these is the Sangamon palaeosol separating the Loveland loess from the overlying Wisconsin beds, and first recognized in Illinois in 1883 (Chamberlin 1883, Voss 1933). The soil covers a wide area and consequently has widely varying characteristics from an accretion-gley in Indiana to a red desert soil in Texas.

The chronological age of a palaeosol includes a statement of both the length of time needed for its development and the approximate date of occurrence both of which, in the absence of absolute dating, are determined by stratigraphical position. There is thus again the danger of circular argument if we use the same (now dated) deposit to infer the age of geomorphological events from the presence of the soil. The deposit preserved as a palaeosol is determined by the intensity of the original soil-forming processes, the time for development and the subsequent geomorphological events. Original zonation of the soil may be effaced by later erosion or leaching so that only a B-horizon rich in calcium carbonate or iron nodules may remain. In these circumstances, interpretation of original and subsequent climatic conditions becomes tortuous. Palaeosols vary from thin dark bands in alluvium, which only superficially resembles a soil, to a complex, polygenetic layer which has suffered several periods of erosion and leaching. It is therefore important that the measurable characteristics of the soil are carefully described to provide as strong a quantitative basis as possible for dating purposes (fig. 2.7).

Two other weathering features which have been used to imply relative age are (i) the degree of decomposition and breakdown of granite fragments, and (ii) the depth of calcium carbonate leaching. Depth of carbonate leaching, for example, was used by Bolton and Worsley (1968) to infer relative ages of tills on either side of the Woore-Bar Hill moraine in Shropshire, and variation of calcium carbonate with depth in a deposit by Leopold *et al.* (1964) in Wyoming (fig. 2.7). To be effective these techniques require a reasonable degree of constancy in both the original composition of the parent materials and the environmental factors which contribute to the leaching of the carbonate, such as slope, drainage and vegetation. The latter considerations apply to weathering on other minerals even where the rock type itself is kept constant, as in Birman's study (1964) of granitic weathering in Sierra Nevadan drifts or in the use of clay minerals to assert equal lengths of weathering and hence of moraine age in the Iberian Mountains of Spain (Thornes 1968).

The use of *tephra* or volcanic dust is similar from a stratigraphic point of view but differs from weathering products in two important respects. The first is that particular eruptions are recognizable by the type of tephra which is disseminated. Secondly, tephra occupies a much more limited range due to the prevailing wind directions and climatic conditions at the time of the eruption. Tephra chronology has principally been employed in Iceland, Japan, New Zealand and the western United States. The smallest stratigraphic unit is that produced by a single explosion but several explosions may occur in a single eruption cycle; thus in a single cycle there may be a sequence of deposits. Nakamura (1960), for example, found in the Tokyo area a basal layer of fallen scoria, followed by lava flows, a layer of ash, reworked ash deposits and finally weathered ash. In Iceland the actual dates of historical eruptions are precisely known, such as the eruption of Oraefajokull in 1362 in south-east Iceland. This makes the tephra horizons very important for studies of local erosion and aggradation rates. For older tephra, especially near the eruptive source, potassium-argon techniques (see p. 43) can be used to provide dates for certain primary constituents of the ash. An important ash bed in the western United States is the Pearlette Ash

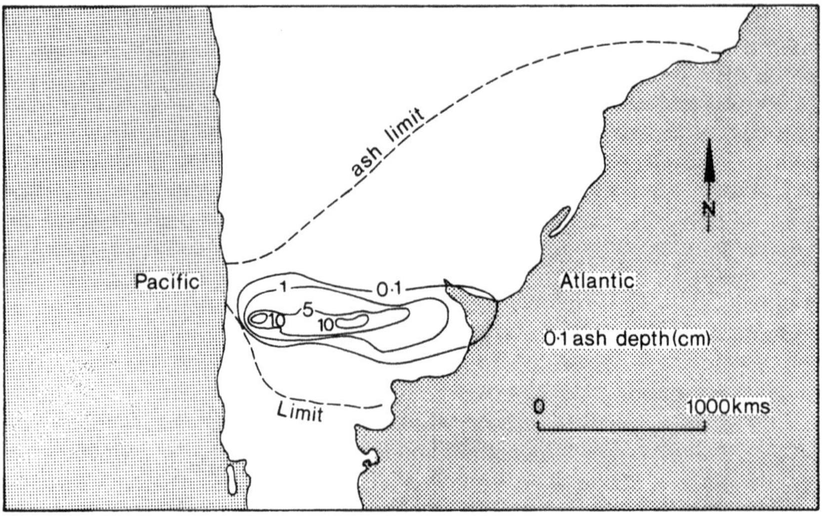

2.8 The ash deposit from the Quizapú volcano in Chile which erupted in 1932 provides an example of how a tephra marker horizon is formed providing a recognizable datum over wide areas (after Larsson 1937). Contours are interpolated between points of equal ash thickness.

which enables the late Kansan continental glacial deposits to be related to the alluvial deposits of the extra-glacial regions and to the alpine glaciers of the Rockies. A similar useful deposit of more recent age is the ash fall from the 1932 eruption of the Quizapú volcano in Chile (fig. 2.8).

Artefactual materials. Artefacts are formally 'man-made objects of prehistoric age', though we shall consider here man-made objects of any age to be of use in determining relative position in time, whether they be eoliths, Roman dams or Confederate bullets at the foot of a Mississippi bluff. Documentary evidence is considered as geochronological rather than stratigraphical.

In the use of prehistoric material we face a difficulty comparable to that associated with the climatic change argument. We must again emphasize the danger of circular reasoning when employing artefacts for dating purposes. Archaeological materials dated at one site by reference to its geomorphological situation should not without care then be used to imply the age of geomorphological events in a neighbouring situation. The problem is of course much less acute because the classification of artefacts has proceeded independently of geomorphological activity. Most of the Quaternary artefacts are tools of flint, quartz and bone. These early artefacts are grouped into tool complexes or assemblages, which represent the cultural development of the people using them. Three qualifications should be made in the use of artefacts: first, because culture boundaries are very sharp different cultural assemblages may represent the same age over short distances. Thus the cultural development represented by the artefactual assemblages of freezer, washing machine and television has sharp boundaries in the world today. Consider, for example, the contemporaneous 'tools' of the occupants of Brazilia and those of an Amazonian village. Secondly, because culture diffuses very rapidly the same artefact may be metachronous in its spatial distribution. The development of neolithic agriculture in the Middle East for example was ahead of that in western Europe. Thirdly, an artefact found on an erosional level only yields a minimum age for the period. Thus, if we find dateable kitchen middens on an upper terrace surface say 8,000 years old, we know only that the surface was in existence at that time — it may have been formed much earlier. There is also the difficulty of incorporation of material. It is frequently difficult to see how material can be deposited in gravels while these are actively aggrading, and yet retain their original character. If they have experienced change, however slight, then they are not *in situ* and could have been reworked. Wartime lookout posts on cliff tops in the 1940s are now frequently found on beaches two or three hundred feet below, often in an upright position!

On the other hand artefactual material is frequently found in association with organic materials such as wood, bone or plant material which provides independent radiometric dates. In western Europe and Africa the tool assemblages seem sufficiently well-known to provide reliable indicators of the approximate age of the sediments in which they are found. Sherds of pottery and glass as well as structures such as water tanks, harbour walls and post holes all provide dates for surfaces and sediment.

Fluvial sediments throughout the world (Zeuner 1952) have been dated very largely using artefactual materials; an example is Miller and Wendorf's study (1958) of the terraces of the Teseque Valley, a tributary of the Rio Grande in New Mexico. Along the Rio Teseque, which heads in the cirques of the Sangre de Cristo mountains, the lowest terrace at 2-3m was dated by radiocarbon and pottery to between AD 1250 and 1880, the period of deposition. The case is of particular interest because rates of sedimentation over the last 2,000 years were obtained using the artefactual material and showed that the average rate of erosion was comparable to that of today. Geochronology frequently employs the reverse procedure, in which rates of operation of processes are used to infer absolute dates (Straw 1964).

The methods discussed here for stratigraphic dating, in the period of greatest geomorphological interest, by no means exhaust the very wide range of techniques available. It is true to say, however, that most of those *not* mentioned are variants of the ones described above. If more extensive discussions of the pitfalls of stratigraphical correlation are required the authors recommend those of Krumbein and Sloss (1963), Straw (1964) or Eicher (1968).

Geochronology

The range of techniques available for absolute dating is much more restricted, the techniques even more specialized and the most reliable extremely costly to execute. They rely for the most part, however, on one simple idea, described by Flint (1970) as follows: 'all estimates and determinations of the date of an event are arrived at by determining (1) the rate of activity of some process in terms of units per year, and (2) the number of such units accomplished by the process since the event occurred; the second is divided by the first'. Thus if one millimetre of sediment is laid down in a year, a bed one metre thick represents 1,000 years.

The problem then is to find processes which are (a) ubiquitous, (b) unvarying through long periods of time, and (c) accurately measurable. Before the advent of radiometric and other relatively precise techniques, much of the dating relied on sedimentation or weathering as the process. These present serious problems either because the rate varies or the number of units accomplished by the process cannot be accurately determined. Certainly contemporary rates of sedimentation in some areas seem particularly high and assumptions based on these rates would give very unrealistic figures. This results from the effect of man-made soil erosion, the over-emphasis on humid mid-latitude data, a maximal contemporary relief, the added effects of Quaternary and Recent eustatic changes and the more 'satisfactory' results achieved by measuring abnormally rapid processes! The rate of peat accumulation was used by G. F. Wright as early as 1881 in dating a kettle hole in Massachusetts, USA, using a rate obtained from northern France! Much the same is true of weathering rates which were used, for example, by Penck and Bruckner (1909) in an attempt to establish the relative lengths of the interglacial periods. In many of these studies the present was an unreliable guide to the past.

There are six basic methods for absolute dating: (1) radiometric determinations, (2) dendrochronology, (3) varve stratigraphy, (4) thermoluminescence, (5) lichenometry, (6) historical records.

Radiometric techniques

Without doubt the greatest advances in absolute dating have come from the development of radiometric techniques which are based on processes involving primary, secondary and induced radio nuclides. These are the decay clock represented by the C^{14} method where C^{14} is compared in ratio to all other carbon in the sample, the ratio clock provided by the differential decay of two radionuclides such as Th^{230}/Th^{232} and the accumulation clock in which a new product accumulates as a result of the decay of an old one as in the potassium-argon technique. The general proposition is that a given radionuclide (P) will decay at a constant rate (λ) to form a product (D). Thus the original number of atoms of a parent (P_0) diminishes such that at time (t) only (P_t) atoms will remain. For example with an exponential decay rate

$$P_t = P_0 \, e^{-\lambda t}$$

and if D is not radioactive $D_t = P_0 - P_t = P_0(1 - e^{-\lambda t})$

or $$= P_t(e^{-\lambda t} - 1)$$

when $$t = \frac{1}{\lambda} \ln \left(1 + \frac{D}{P}\right)$$

Radio carbon (C^{14}) is present in the atmosphere in an equilibrium state so that the quantity remains relatively uniform. While a plant or animal is alive, it maintains a fairly constant level of C^{14} which is in balance with that in the atmosphere through the plant's metabolism. Upon death the C^{14} begins to disintegrate at a known rate (a half life of 5730±40 years). It is this known rate of disintegration which makes the nuclide a useful timekeeper. Unfortunately it disintegrates rather rapidly (i.e. it has a relatively short half-life) and so it is useful for a limited dating period of approximately 50,000 years. Moreover, uncertainty in measurement increases with the age of the sample so that 30,000 years is often regarded as the practical limit and 2,000 — 20,000 years the most reliable limits.

Debate as to the accuracy of the technique revolve around (a) the half-life to be used, (b) variations in the concentration of atmospheric C^{14}, and (c) contamination of the sample. In 1962 a Cambridge conference revised the half-life of radio carbon from 5,570 ± 30 years to 5,730 ± 40 years; this necessitated readjustment of some of the dates determined earlier. Variations in the C^{14} content of the atmosphere can be caused by changes in the cosmic-ray intensity (equilibrium is assumed for dating), the magnitude (but not orientation) of the earth's magnetic field and the degree of oceanic mixing, because the ocean acts as a buffer for CO_2 in the atmosphere. New dendrochronological dating scales have been devised, using the Bristlecone Pine. If these are taken as being absolute and correct, compensations can now be made to C^{14} dates to resolve some of the apparent discrepancies between C^{14} dates and known historic events. It also seems that in oceanic and fresh-water systems adjustment has to be made for the apparent age of the water. This is quite important because both foraminifera and mollusca have been extensively used for radiometric determinations. Variation may also be caused by the effect of percolating ground water, the incorporation of old carbon, contamination in the field, in the laboratory and during storage. A 1% contamination with foreign 'modern' material will, according to Olsson (1968), result in a measured age which is 10,000 years too low for a 44,000 year-old sample, and 1,000 years too low for a 22,000-year-old sample. On particular materials, such as lacustrine carbonates, finite C^{14} dates seem rather unreliable beyond about 22,000 years. Compensation must also be made for contemporary increases in atmospheric carbon (fuels) which lead to errors of approximately 80 years for each 1% estimate error.

The ratio method, for example comparing Protactinium^{231} with Thorium^{230}, has been used particularly for dating deep sea cores, especially since this ratio is not too sensitive to changing oceanic sedimentation conditions. The decay constants differ by a factor of two and thus the ratio changes through time in a regular manner. It is thus a useful method though valid only for ages up to 150,000 years. Potassium-argon dating, the third process type, can span the whole of the Quaternary and latest Tertiary and therefore should offer, as more dates become available, considerable aid in the study of denudation rates. The technique is notably restricted to igneous rocks containing primary biotite, muscovite or amphibole and it is very sensitive to slight amounts of rock detritus in the sample. Since 1907 when the disintegration of uranium to lead was first used in dating there has been a rapid advance in such techniques. Today, U^{238}, U^{235}, Th^{230}, Th^{232}, Rb^{87}, K^{40}, Ra^{226} and H^3 (for periods of less than 100 years) are all used with success; in many cases with an error of less than 2%.

Dendrochronology

Dendrochronology (Douglass, 1914), a technique which covers a much shorter period than most radiometric techniques (about 7000 BP) is so accurate that it has been used to calibrate adjustments necessary to radiocarbon dates. This technique may be defined as 'the systematic study of tree rings applied to dating past events and evaluating climatic history' (Fritts 1966). It can be applied with varying degrees of success wherever annual rings grow on trees, but in addition it has been used to infer, along climatic margins such as semi-arid lands and in areas where temperature limits growth, the changes in past climates.

Each year as a tree grows, the xylem adds a ring of growth which includes inner and outer, early wood and late wood growths with a transition between them (fig. 2.9). Straightforward application of the technique or the rate principle, discussed at the beginning of this section, relies on the clear development of one ring each year and on some control on ring thickness variability, usually a climatic cause of aridity, humidity or temperature which affects the relative availability of soil moisture to the tree. It is perhaps unfortunate, in examples near glacial margins where knowledge of past climates can tell us most about the geomorphological history of the region, that local variations in soil moisture, which inhibits the growth of the cambial layer, reduces the use of the technique so that more elaborate dating procedures may have to be used.

The simplest procedure involves the measurement of the ring sequences on the radii of a specimen. The relative thickness of these sequences is used as the basis of correlation by overlapping the plots and matching the peaks and troughs of the graphs (fig. 2.9). From many specimens, standardized, composite and eventually regional master diagrams are produced starting with modern trees and progressing to older, perhaps fragmentary remains. This technique of cross-dating involves the assumption that in any area of climatic uniformity the ring patterns through time will be the same in all trees in that region. Because trees overlap in age, and therefore in 'tree calendars', the series can gradually be extended further and further back in time. It is also possible to establish what are called 'floating chronologies' in which sets of tree ring series may be correlated without being tied to a known date. Hopefully, this may eventually be tied in with a regional master chronology.

As with all techniques for dating, several difficulties arise in interpretation of the results. Notably, the effect of the time patterns of precipitation during the growing season on ring thickness; the differential growth rates of tree ring

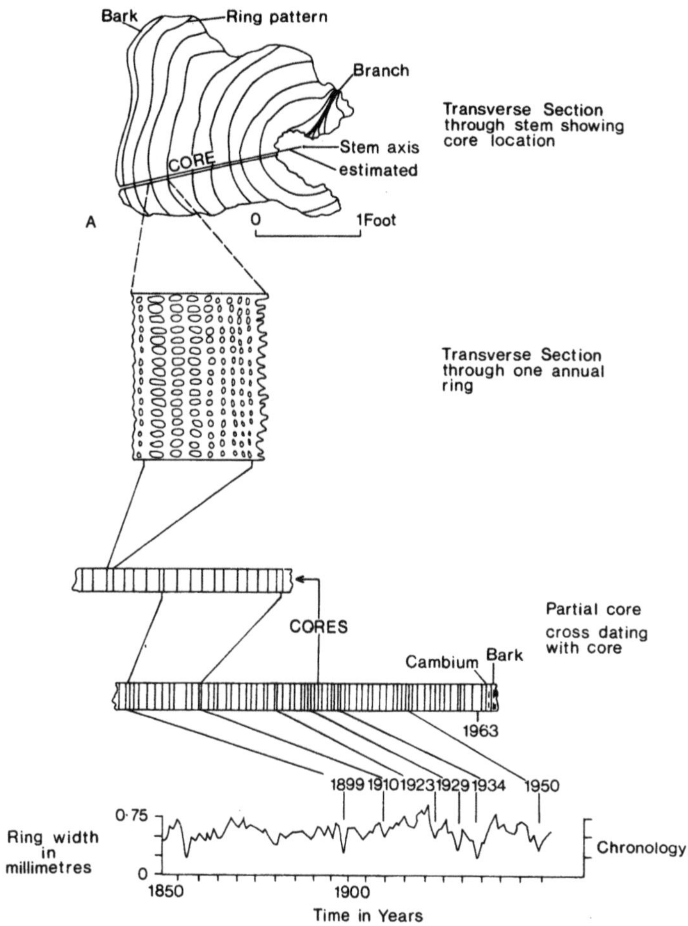

2.9 Dendrochronology. A core taken from the bark of a tree is used to establish a pattern of tree ring variability. Cross dating with patterns from other tree and floating chronologies allows the record to be extended backward from the pattern in a present tree (after La Marche Jr 1968).

with increasing tree age; the movement of the youthful zone of rapid growth, up the tree with age, so that ring variability changes along the tree length, and finally local details of tree-to-tree and site-to-site variability.

It is implicit that wood fragments must be available and adequately preserved, a fact which favours the use of this technique in some particular areas; as for example in the south-west of the United States where a continuous series of rings from 59 BC to the present has been established. The technique is also important for geomorphological process studies when applied to existing trees. Sigafoos and Hendricks (1961) applied the dating technique to evaluate the recent behaviour of the Nisqually Glacier on Mount Rainier (Washington State, USA). Here they were able to locate the maximum recent downvalley advance of the glacier by the sharp change in size and age of trees. The age of a tree is interpreted as the minimum period that has elapsed since the ice left the position occupied by the tree. The minimum age of a land-surface such as a moraine is then the age of the oldest tree on that surface. In such cases, however, it is important to establish the regeneration time for vegetation to establish itself on the new surface. Trees may grow on surfaces which are not completely stabilized and occasionally they are found up to 100 years old on thick supraglacial deposits. The time since active scouring may therefore be longer than the minimum mentioned above. In this kind of inquiry, the availability of tree seeds is also of importance because it is a factor in the rate at which the forest increases in density. Finally the rate of growth of the trees in part determines their suitability for dating, the slower-growing trees being preferred. A recessional sequence for the Emmons Glacier (after Sigafoos and Hendricks 1961) is shown in fig. 2.10. By plotting the thickness of tree rings through time, it has been possible to use statistical techniques, and particularly cross-spectral

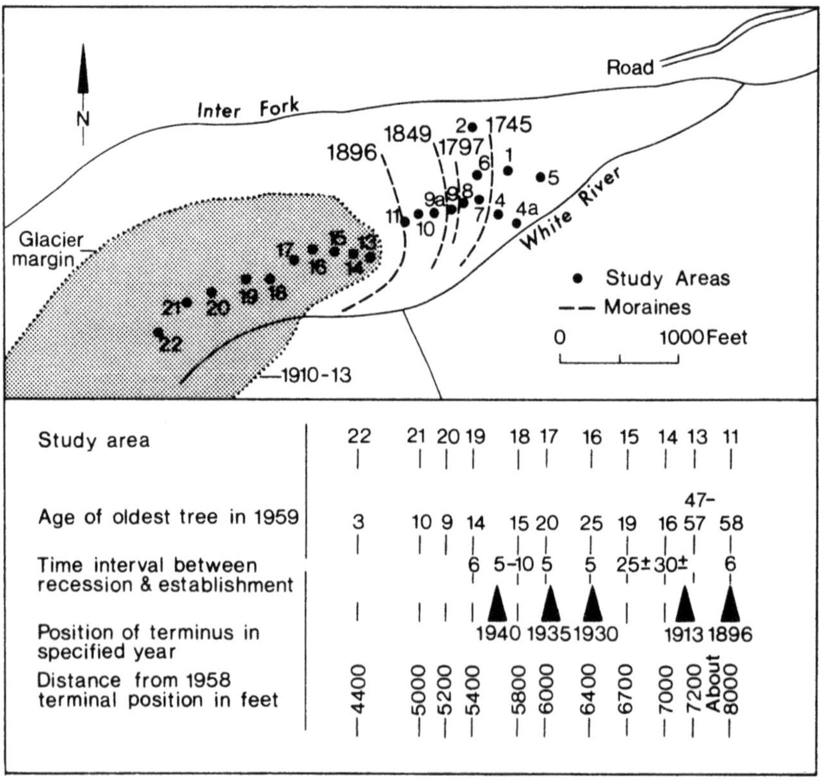

2.10 Recessional sequence of the Emmons Glacier based on dendrochronology (after Sigafoos and Hendricks 1961).

analysis to compare different series for their degree of similarities and differences — a technique we shall discuss at some length in chapter 4. This procedure has also been applied with considerable success in varve chronology.

Varves

Varve analysis is one of the oldest methods employed for absolute age determination and has been applied mainly to dating Quaternary, as opposed

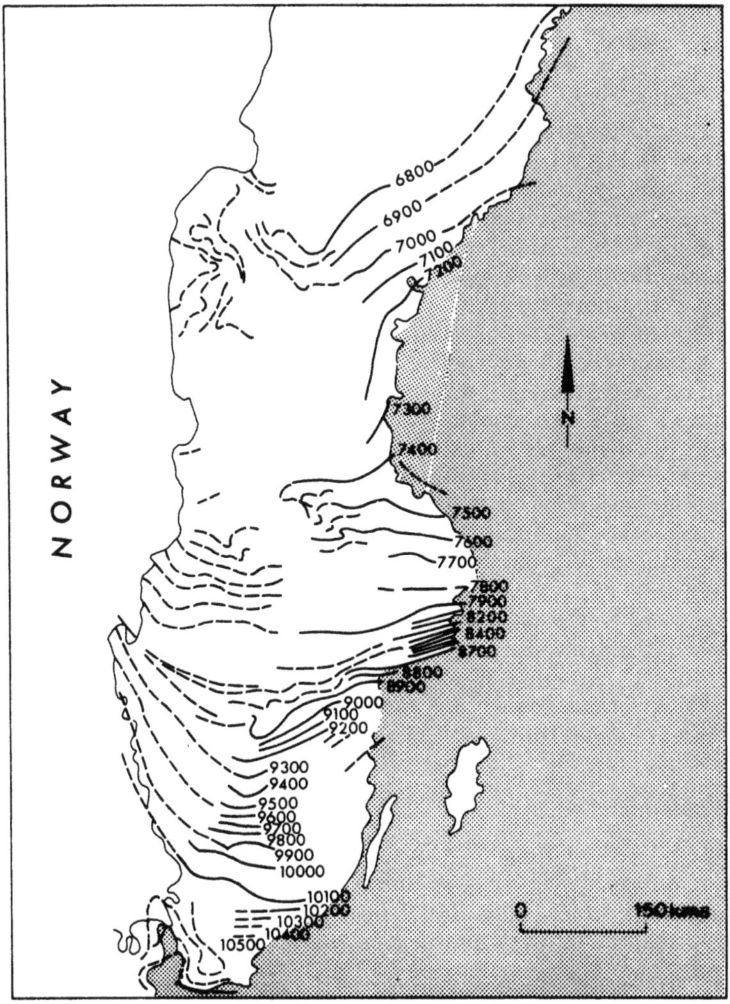

2.11 Chronology for stages of retreat of continental ice from Sweden based on de Geer's analysis of varves (after Rankama 1965).

to Recent, geomorphological events. Varves have been particularly useful for short distance correlation. They are essentially lacustrine deposits formed in rhythmic couplets in which pairs of lithologically dissimilar strata are laid down annually. Ordinarily these consist of alternating silts and clays deposited in pro-glacial lakes, ranging in thickness from a few millimetres to a few centimetres. The clays are believed to separate out from the silts because of the delaying factor induced by increased water density resulting from thermal changes and by rapid deposition of coarser material in the spring and summer melts. Most studies are based on the assumption that the varve is annual and, as with dendrochronology, attempts have been made to set up regional master chronologies. De Geer's chronology (1912) for southern Sweden based on localities only a kilometre apart, but spreading over a 1,000 kms, is one of the best known and has provided dating for 12,000 years from AD 1900 with only one gap in the sequence (fig. 2.11). It is compatible with C^{14} determinations near the middle of series and the results have been confirmed by several workers.

The success of dating in Scandinavia is partially attributable to the fortunate circumstances in which the Baltic Ice Lake received sediments which could be well-preserved and which were distributed uniformly over a wide area. The North American attempts led by Antevs (1953, 1955) for 950 km of the Connecticut Valley have met with much less success due to gaps in the sequence, and there is in the chronology for Canada a discrepancy of many years between Antevs' chronology and current estimates based on C^{14}. One American chronologist was thus led to conclude that 'varve analysis does not appear to be a reliable means of long-distance correlation' (for an extended discussion see Flint 1970).

Certainly there are problems as yet unresolved. There are no definite criteria for considering all of the laminations to be of annual rhythm and we still have a somewhat limited knowledge of the process of varve formation. In Scandinavia some geologists consider that disturbance of sediments by storms would lead to redeposition in thinner layers and a false sequence. Finally, there is the difficulty of long distance correlation. This is only possible if there has been a similar history of temperature changes and deglaciation in two areas. It was on this basis that Antevs tied Scandinavian to North American sequences. In general, however, the technique is best restricted to local histories.

Thermoluminescence

A method based on the same principles but still somewhat experimental and applying only to short periods of time is thermoluminescence. This is used for dating pottery which, when found in association with hillslope or fluvial deposits, may be an important factor establishing the rate of operation of processes. The procedure is based on the fact that all pottery and ceramics contain radioactive impurities which emit alpha particles at a known rate depending upon their concentration in the sample. These in turn create a source of stored energy in the pottery minerals which are stable at normal temperature. If the material is heated, the trapped energy is released in the form of light. When the pottery is fired, all the energy emitted and stored up to that point is released; then the storage process starts again. Methods have been devised for measuring this stored energy and, after allowing for the initial radioactive content and the susceptibility of the pottery material to produce light under controlled conditions, an absolute age may be derived. Hall (1969) estimates a possible error of ±10—15%, but states that in good circumstances the date may be obtained to ±3%.

The amount of thermoluminescence is also related to the constancy of temperature which influences the proportion of electrons trapped or escaping from the crystal lattice of the minerals involved. The leakage rate is a function of the given constant temperature. Variation in maximum temperatures lead to a greater or lower rate of release. This principle has been used to date the period when the constant temperature of an environment underwent a change. It is of restricted use in natural circumstances since there are few geomorphological environments with a constant temperature but it has been used in studies of cave environments in comparison to similar materials outside the cave (Ronca and Zeller 1965) and potentially has a use in comparing surface and subsurface regolith dates.

Lichenometry

Lichenometry uses, to establish rate per annum, the radial growth of certain species of lichens. The method applies particularly to recently exposed rock surfaces or recently active geological processes in treeless areas. Different lichens grow at different rates and none grow at constant rates throughout their entire life. Early growth rates are microscopic and for a long time the thallus remains invisible, then there follow several decades of rapid growth which eventually slows down to a rate which remains constant over centuries. Growth rate curves for species are obtained by comparing diameters on known surfaces, such as moraines determined from aerial photographs. An example is shown in fig. 2.12 for the growth rate of the most commonly used lichen, *Rhizocarpon geographicum.* The principle problems relate to lack of information on the nature of lichen growth rates and the effects of local environment on the growth. Most workers set up a regional growth rate by making use of other dating techniques, and then use the curve to establish within-region differences. The most important local environmental factors in the Central Alaska Range were found by Réger and Péwé (1969) to be stability of the substratum, sunlight and moisture. The growth rate in this area was found to be almost parallel to that established in the Steingletscher of the Swiss Alps by Beschel (1961). Since Beschel's classic work the technique has been widely applied in northern Canada (Andrews and Webber 1964, 1969), the Rockie Mountains in Colorado (Benedict 1967) and Scandinavia (Stork 1963, Jochimsen 1966, Andersen and Sollid 1971, Mottershead and White 1972, and Matthews 1974). The technique is thus shown to be of great use in areas of recent origin and particularly near ice sheets (fig. 2.13). The principle problems relate to the establishment of

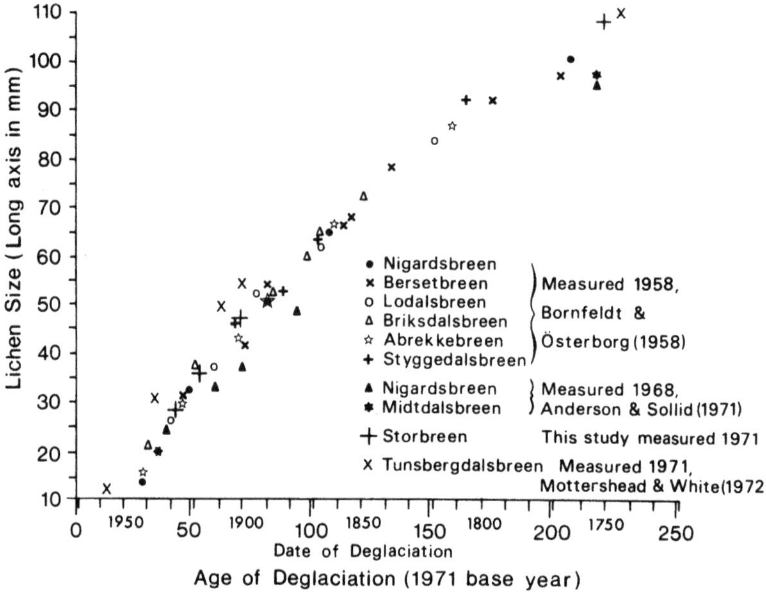

2.12 Regional growth rate curves for *Rhizocarpon geographicum* for different Scandinavian locations based on published literature and collected by Matthews (1974).

regeneration times and growth rates, the effect of local variations in aspect and moisture and the problem of species survival from disruptive events (glaciations) or impact by transport from other areas. Measurement difficulties include the choice of the largest lichen in the sample plot or mean values of say the five largest lichens present. Benedict (1967) and Matthews (1974) provide a full discussion of these problems.

Historical records

Finally, attention should be drawn to the use of historical records to establish a time scale for a small area. Such information, which is becoming increasingly important as our records improve, includes casual observations of catastrophic events, natural history diaries, explorers' or military records, descriptions, prints, photographs and maps. The measurements record the instantaneous state of a situation but not the intervening processes but are valuable in providing an unambiguous date for an event from which other geomorphological events can be dated.

The information provides a time scale in two ways. First, it may be used to provide a precise date for a deposit formed at some time in the past and now preserved in the stratigraphy of the area. An ash layer resulting from a recorded forest fire, or a mud layer from a dated landslide or flood serve as

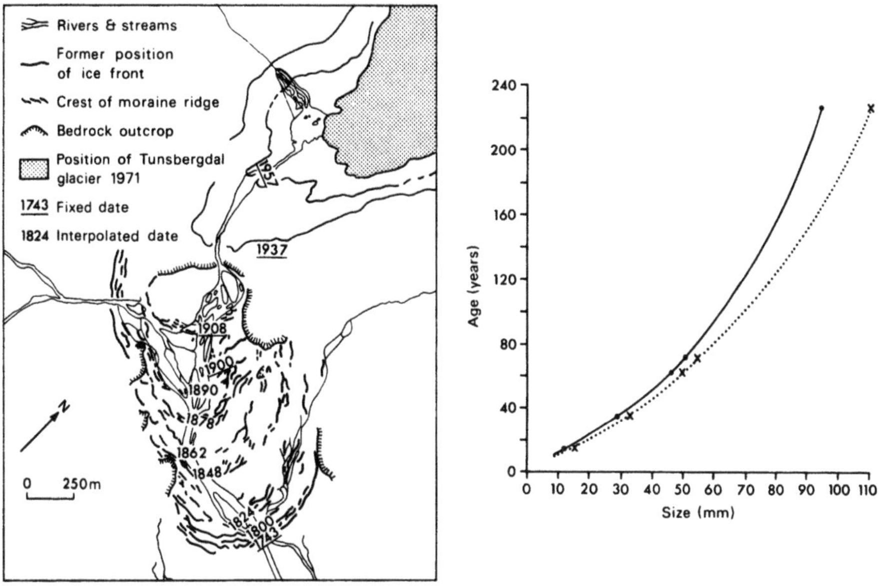

2.13 The use of *Rhizocarpon geographicum* in the Tunsbergdal valley, Norway, for dating moraine sequences. In the upper diagram the dated sequence is shown, in the lower diagram the plot of age against size for the same location. The solid line is based on the largest five lichens per plot. The dotted lines uses the largest lichen diameter Zafter Mottershead and White 1972).

examples. The technique involved is essentially stratigraphical and similar to marked horizons in the geological record from which other events are placed in a time scale. Where historical records exist, however, the date is usually very precise.

The second technique involves the use of the visual record as an instantaneous picture of an area against which the present landscape may be compared. This is particularly valuable for vertical air photographs which may be used to provide a time base for intervening or subsequent events. Such techniques are, however, of most use in the measurement of variables in time or for establishing rates of operation of processes and are considered more fully in the next chapter.

3 The measurement of variables in time

The measurement of variables such as cliff recession, temperature or soil movement in an already established temporal framework involves such issues as the method of sampling in time, including regular or irregular spacing of records, the frequency of observation and the types of 'observational clocks' available against which to standardize the information.

Incidental observation

We may describe the crudest level of measurement as incidental — an event is recorded merely because it occurred or because it was regarded as unusual. In most cases such observations concern only the extreme event which is usually one which caused damage to life or property. The terms 'catastrophic' or 'disastrous' clearly specify this type of event and record. Examples range from earthquakes, hurricanes, tornadoes and floods to famous or infamous landslides or 'exceptionally high' tide levels.

Usually, unless the likelihood of the event reoccurring stimulates government research and remedial works, as occurred with the 1953 North Sea floods, the event passes into local history and the information is imprecise and insufficient for detailed research. Occasionally the record is a thorough one, due either to the presence of an interested scientist in the neighbourhood or because the event was recorded on pre-existing continuously recording stations. An excellent example concerns the famous landslide ofChristmas Eve 1839 at Bindon on the east Devon coast. Here, due to the fortuitous presence of two geologists Conybeare and Buckland (1840) and an excellent artist, Mary Buckland, an immediate and thorough study was made and their account is still quoted in current studies. The fact that the event was accompanied by a 'wonderful crack', and a sound like 'the rending of cloth', a 'heaving of the beach' and 'flashes of fire and a strong smell of sulper' with attendant rumours of devils and volcanoes, only served to enhance interest in this 'most extraordinary and terrific explosion of nature' and thereby ensure that interest, drawings, photographs and visits by geomorphologists would be made for many years to come. There was even a piece of popular music 'The Landslip Quadrille' to celebrate its occurrence,

3.1 The Landslip Quadrille. The cover page of a piece of music written by Ricardo Linter and drawn by Daniel Dunster to celebrate the famous 1839 landslide at Bindon, Devon.

surely one of the few geomorphological events since the Noachian deluge to be so celebrated (fig. 3.1). From such records interesting and amusing information can be gleaned but in the majority of cases the surveys are inadequate for precise research purposes. For events which occur in uninhabited areas or which cause little damage to property the record is always poor.

A second type of incidental measurement includes those made at a fixed time but without consideration of the frequency of observation required for the surveyed subject and often for a purpose other than that for which the data is finally employed by the geomorphologist. This is the case with historical map data which has not been collected on a routine basis. Examples are Steers' (1948) illustration of the changes occurring on the north Norfolk coast from Hunstanton to Brancaster between 1883-5 and 1935; de Boer's (1964) summary of the growth of Spurn Head or Kidson's (1952) surveys of erosion on Dawlish Warren (fig. 3.2). Aerial photographs are also proving of value now that repetitive cover is available (fig. 3.3) and these,

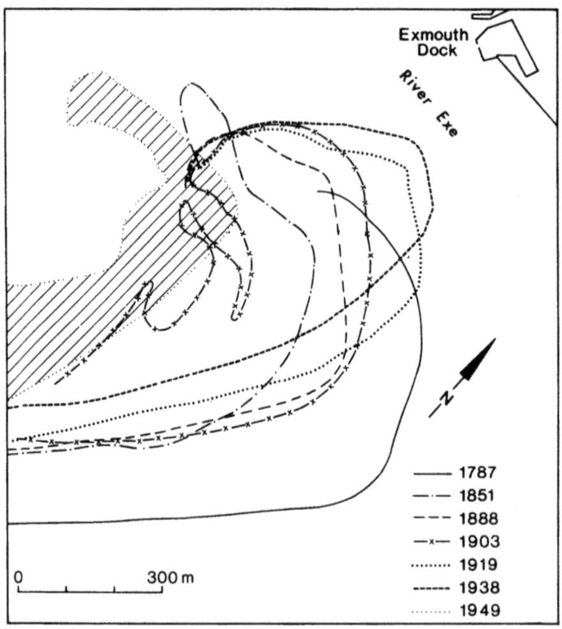

3.2 Changes in spit outline on Dawlish Warren at the mouth of the river Exe. Selected outlines from historic surveys (summarized in Kidson 1952).

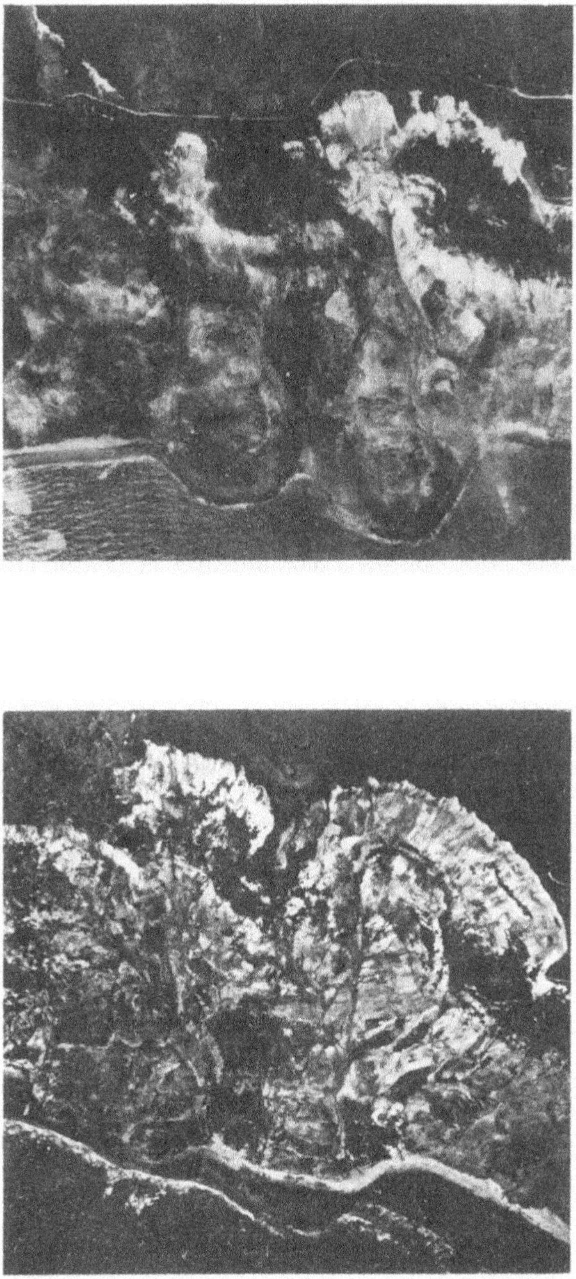

3.3 Air photographs of Black Ven landslide in 1958 and 1969 to illustrate the increasing importance of repetitive cover in historical geomorphological studies and surveys of rates of erosion.

together with the accurate plans of the national surveys, extend the reliable record. In Britain aerial photographs extend back to 1940 with Ordnance Survey maps to approximately 1900. Carr (1962) has warned of the dangers of accepting uncritically the sometimes doubtful accuracy of the early surveys and additional dangers lie in the interpretation of the record, for it is always difficult to infer process or rates from a series of 'spot' pictures. Geomorphologists are, however, fortunate to have such information for they yield, however tentatively, information on the rate of operation of processes for a much longer time period than any we have since observed formally on a regular basis. Incidental aerial photographs for military or civil (but non-geomorphological) purposes are very precise but like historical maps are unhappily relatively limited in process and evolutionary studies for the simple reason that rates of operation have to be 'smoothed' or averaged between the observations and the period between observations is variable. Trivially, we may compare this with the experimental geomorphologist who observes his laboratory experiment in between and incidental to doing other things.

Map records are sometimes supplemented by literal descriptions, prints, oblique photographs or diaries of explorations or military expeditions. The works of Powell (1875) on the Colorado River or Dutton (1880-1) in the American west are classics of this kind as is the natural history diary of Guthrie-Smith (1926) at Tutira in New Zealand or Darwin's accounts of the voyage of the Beagle (1839). Later these records may be used to build up a precise story of the occurrence of geomorphological events as demonstrated by Bryan's (1923) studies of soil erosion in the American south-west which utilized military records with brilliant success.

Finally, there are a few measured records which because of their longevity are very valuable. A few such as the records of the flood stages of the Nile, AD 622 to the present, (Hurst 1950) or the level of the Caspian Sea, 1839 to the present, (Leopold *et al.* 1964) are of considerable length but the majority of useful records rarely extend beyond 100 years. The longest concern fluvial runoff and tides and it is now of vital concern to geomorphogeomorphology to maintain recording stations for other processes. This is particularly true now that reliable automatic recording devices are readily available.

It is a strange fact, however, that many of the most extensive records remain essentially unanalysed. Much remains to be done on the magnificent fluvial and sediment figures collected by the United States Corps of Engineers for the Mississippi river or the similar lists preserved in hydrological institutes all over the world. An urgent task is to develop data banks and analytical systems to cope with the current explosion in data collection.

Controlled observation

A second order of observation in time is that which may be regular, but suboptimal in terms either of the seasonality of the process under observation or of its irregularity. A very common but unavoidable example of this is that, for logistic reasons, observations on high-arctic processes may only take place continuously in the summer months because food and supplies cannot be brought in or made available in the winter. Our understanding of summer-season processes in these and comparable logistically-bound regions is much better than our knowledge of what happens in winter. In a remarkable investigation of activity in front of the Biafo glacier in the Karakoram Mountains, throughout the winter as well as summer seasons, Hewitt (1967) showed quite clearly the significant differences in processes in the two seasons.

The second problem, that of irregular events and regular observations, presents one of the biggest problems for geomorphological investigation. Continuous observation is an expensive business and yet the *only* real way to catch the rare (irregular) but sometimes most important event. How often one hears the researcher (and no less the undergraduate) complain that after weeks of observation 'nothing happened' only to learn that, the day after his departure, a flood caused unprecedented erosion and channel changes! The basic irony is that we need a lower density of sampling for regular events than for the very infrequent ones where a high density is very desirable. Unfortunately in practice the reverse is usually all that we can afford.

It comes as a surprise that, after fifteen or twenty years of quantitative and theoretical work in geomorphology, little attention has been given to the basic problem of experimental design for time sampling. This is especially true when so much has been done for spatial experimental design (table 3.1). Much more seems to be known about the analysis of temporal geomorphological data than its collection.

A few attempts are being made to obtain data in pilot studies which will provide better sampling schemes for future investigations. As an example, McCann, Howarth and Cogley's paper (1972) sets out to assess the variations in hydrological characteristics of a small Arctic basin with a longer term project in mind. Here one of the difficulties mentioned above was encountered; it proved impossible to install water level recorders until after the peak flood because the creeks were completely covered with snow overlying ice in the stream. The authors were, however, able to make measurements in the early stages of flow and during peak flow using area-velocity techniques. A result of these technical and logistic problems meant that sections of the hydrograph had to be interpolated and some gaps occurred when it was impossible to read the hydrographs. In this example, a continuous phenomenon could only be observed at regular intervals for irregular blocks of time because of unforeseeable difficulties (fig. 3.4a). A second example is Gardner's study (1969) of superficial talus movement in the Lake Louise area of the Canadian Rockies, apart from which there is almost no other information on this important Alpine process. The data is for three years, 1965-7, during which absolute amount and pattern of surface movement were observed. Part of the difficulty here is that there is a variety of types of movement; true creep and subsidence may be virtually *continuous*, rolling and sliding *discontinuous* and almost certainly seasonal, and the effects of avalanching undoubtedly *seasonal.* Secondly, for logistic

Table 3.1 Some common types of measurement in time

Type	*Description*	*Time period*	*Data type*	*Nature of event*
Incidental	(a) Event record (b) Cumulative change	Irregular	Historical record; documents, maps, air photographs	Unusual flood, landslide
Record made on occurrence	'Expected' event recorded when it occurs	Irregular	Conscious recording designed to catch events of certain magnitude or type	Expected flood, e.g. annual flood of Nile
Long interval a) Annual b) Seasonal	Cumulative change since the first visit on annual or seasonal basis	Regular period	Designed experiment	e.g. Annual glacier movement e.g. fluvial activity in seasonal environ-ment-arctic-arid.
Short interval	(a) Cumulative change since last visit (b) Continuous measurement *on* day of visit e.g. short time sample once a month	Regular period	Designed experiment	e.g. Mudflow movement: trace experiments on beaches; resurveys of beach profiles.
Pilot study preceding long term study	(a) Short term continuous sample to design time sampling frequency (b) Regular measurement for irregular blocks of time	1. Continuous 2. Spot sample	Sample and measurement design for information content studies	Any continuously acting phenomena — should be applied to all geomor-phological process studies
Long term	Continuous record	Permanent station	Complete record with full experiment design	Any variable process with irregular events or continuous flow character
Long term but with temporal sampling design	(a) Regular (b) Nested time sample (c) Random time sample (d) Frequency determined by variability of process and logistics	Time period chosen to yield maximum information on variation	Sample design based on pilot study	Any variable process

reasons the observation had to be made at twelve-month intervals with the result that accumulated movement is being measured over a wide variety of possible movement conditions. To know what factors were contributing to variations in amount of movement a much denser period of observation would be required, but of course throughout most of the year the talus is snow covered. A rapid, irregular and in some cases considerable downslope shift of surface material probably occurs over a short period but this cannot be established from the data (table 3.2). Brunsden has worked on the rate of movement of mudslides in Dorset; here a regular scheme of measurement was set up on a monthly basis to assess the longer term (three-year) variations in monthly cumulated movement. The principle problem here is that movement is irregular in time and very rapid in occurrence. Under the logistic circumstances regularly spaced observations as close as possible were necessary; apart from continuous recording there appears to be no alternative procedure. In addition, the logistic problems for the measurement day itself meant that only the cumulative distances moved in the preceding

Table 3.2 The movement of talus material in the Lake Louise area recording the net result of several different processes (after Gardner 1969)

Slope	*Transect*	*Observation period (yrs)*	*No. of particles*	*Maximum movement (m)*	*Minimum movement (m)*	*Mean movement (m/yr)*	*Particles with no movement (%)*
A	1	2	21	45.4	0	3.48	15
	2	2	50	70.9	0	3.53	12
	3	2	61	29.3	0	1.88	2
B	1	1	25	.4	0	.06	64
	2	1	35	.3	0	.01	94
	3	1	59	.5	0	.04	82
C	1	2	31	6.9	0	.36	66
	2	2	41	1.2	0	.04	59
	3	2	47	2.8	0	.09	51
D	1	2	33	1.0	0	.18	24
	2	2	46	1.9	0	.14	61
	3	2	39	1.3	0	.13	49
E	1	1	14	41.1	0	3.78	29
	2	1	29	4.1	0	.71	31
	3	1	30	4.7	0	.51	40
F	1	1	18	2.7	0	.43	39
	2	1	14	.5	0	.17	38
G	1	1	31	1.1	0	1.24	17
	2	1	12	7.5	0	1.43	33
	3	1	22	1.7	0	.33	59
H	1	1	23	18.7	0	7.14	13
	2	1	25	37.0	0	9.06	16
	3	1	16	25.4	0	5.89	31
I	1	2	36	31.2	.2	2.01	0
	2	2	35	60.0	0	1.55	29
	3	2	26	7.6	0	.46	27

month could be recorded. It would have been preferable to measure the rate of movement on the day of visit itself so that the relationship between movement and other variables recorded on that day could be examined but a suitable surveying precision was not possible in the time available. The rates, as far as they can be evaluated from monthly data are shown in fig. 3.4b.

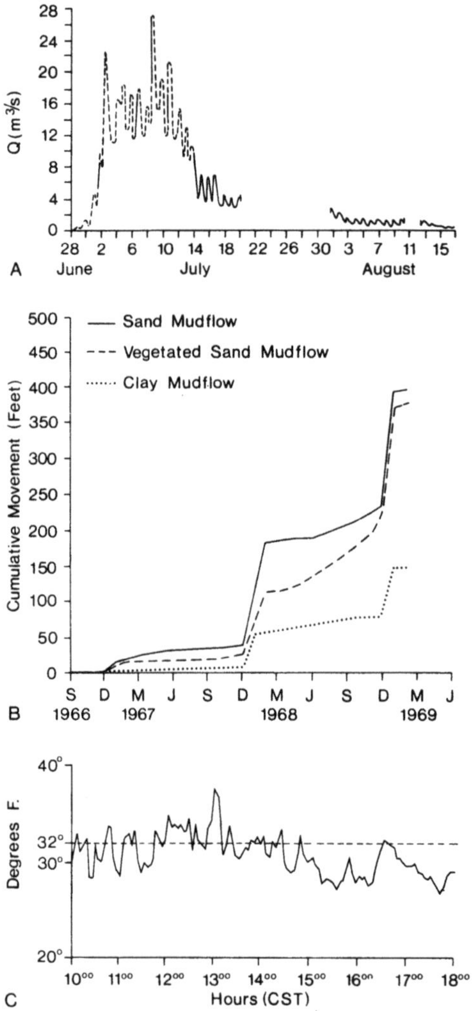

3.4 (a) Discharge record for the river Mecham, Queen Elizabeth Islands, 1970. Dashed sections are interpreted on the basis of qualitative observations between measurements determined by area/velocity techniques and later discontinuous stage recorder records (after McCann, Howarth and Cogley 1972). (b) Cumulative movement on three mudflows in Dorset, UK, based on monthly measurement of cumulative movement of surface stakes. (c) Soil temperatures at Resolute Bay using continuous recording to demonstrate diurnal variations (Cook and Raiche 1962).

The final example is the analysis by Cook and Raiche (1962) of ground surface and soil temperature at Resolute Bay in the Arctic permafrost region. Here continuous records are used to demonstrate very regular diurnal and seasonal variations. Where very regular events of this type exist refined sampling designs, taking account of such regularity, can yield a great deal of information with a minimum of observation. Observation in time under these circumstances is at its most efficient (fig. 3.4c).

The examples chosen illustrate some typical problems; they are of an exploratory nature and in all cases the authors are fully aware of the problems involved. A survey of the literature, however, suggests that time sampling considerations have not, by and large, been given the prominence which they deserve; some authors do not even supply information on the rate and frequency of observation.

Some designs for temporal sampling

The principle types of time base for geomorphological events and the associated sampling problems and likely designs may be outlined as follows (fig. 3.5):

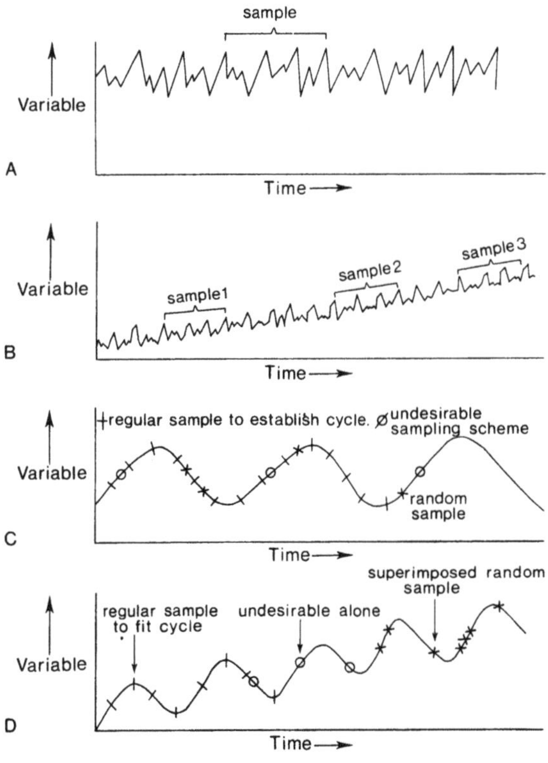

3.5 Sampling schemes for various types of temporal data. (a) Stationary. (b) Trending. (c) Cyclic. (d) Trending and cyclic.

(1) Situations in which there is no variation in the mean and variance through time and the variables of interest have a high or low variance. This kind of geomorphological process is considered stationary. The background variation is called noise. Provided that the noise can be characterized, a single sample which determines the mean value adequately specifies the values taken on by the processes. Geomorphological processes of this type are relatively rare except on a very short time scale (fig. 3.5a).
(2) As scales change or if a long term trend is present one kind of sampling situation passes into another. Thus river channel depth at a section may remain steady for several months or even years, while in the longer time span it may gradually deepen. In this case widely spaced discrete observations at annual intervals would suffice to describe the trend (fig. 3.5b), and the steady state fluctuation.
(3) Situations in which there is a regular cycle of the variable in time are rather more common. If the noise is low, the curve regular and trend absent at the scale of interest (fig. 3.5c) then the process can be characterized by a sample which is adequate to specify the typical wave length and amplitude. Tide cycles, temperature cycles and seasonal variations in discharge may take this form. The sampling interval in regularly varying situations depends on the degree of precision required in fixing the shape of the curve. For example, diurnal cycles might be fixed by an hourly sample, weekly fluctuations by a daily sample and seasonal changes by a weekly or monthly sample. A problem may arise if the frequency of regular observation corresponds with some frequency or harmonic in the variable. It is therefore sometimes desirable to include a superimposed random sampling scheme on the systematic observations. A general rule is to sample at a frequency lower than that required to determine the harmonics of the oscillation.
(4) If, in the long term, trend is present in addition to the regular cycle it can also be determined if the frequency of observation is lower than the harmonics of the cycle (fig. 3.5d) either by a long period during which samples are made or by several samples in time each long enough to specify the cycle. For example, trends in annual mean temperature or precipitation may be superimposed on the basic seasonal rhythm. This may also be true as a system settles down towards a new equilibrium after a shock, as in the long term sediment yield of a basin recently devegetated or disturbed by a major flood.
(5) Situations in which events occur with some degree of regularity in time but whose arrival can only be stated probabilistically. For example, a discharge of 1,000 cumecs might occur on average once a year but the likely location of the event (for example in the 'rainy' season) is not known. Many geomorphological processes seem to have this characteristic of exhibiting a certain regularity which is inadequately specified. The data may include irregular cycles and trends which are usually hidden by the variance of the process. River runoff, landslide activity and soil movement are probably of this character. There is only one possibility here — to observe as frequently as possible (which ultimately means to observe continuously).
(6) Extreme events by their very nature cannot be designed for in sampling,

although for engineering or the construction of sampling equipment a reasoned guess must be made. Such events are captured accidentally in a pre-existing framework of observation or reported incidentally, e.g. landslides.

Recording geomorphological events

The recording of events in time depends on one of three principles: (i) a continuous trace, for example of river stage, on a graph against time — here the graph is operated by a variety of clockwork and electronic mechanisms (fig. 3.6); (ii) periodic observations of the variable at predetermined intervals by an observer, which is the most commonly employed technique in geomorphology; for example, we may observe either the rate of movement during say twelve hours on the recording day and in this way build up a sample of rates from all possible daily rates, or we may observe the cumulative activity that has occurred since our last visit, for example the amount of loss from pins on a cliff face (fig. 5.3 and p. 105); (iii) a system in which the time is recorded when an event takes place. This system is employed in the tipping rain-gauge bucket principle (fig. 3.7). A clock operates continuously and a mark is registered every time a bucket, which fills with water, tips. Variations on this theme are also very common in current process investigations. The principle problem here is the reset time of the recording device. If rare events occur in quick succession more rapidly than the instrument can respond then information is lost. The first and last of these three basic methods can readily (though expensively) be linked to various multichannel recording systems which store the information and repeat it as and when it is required.

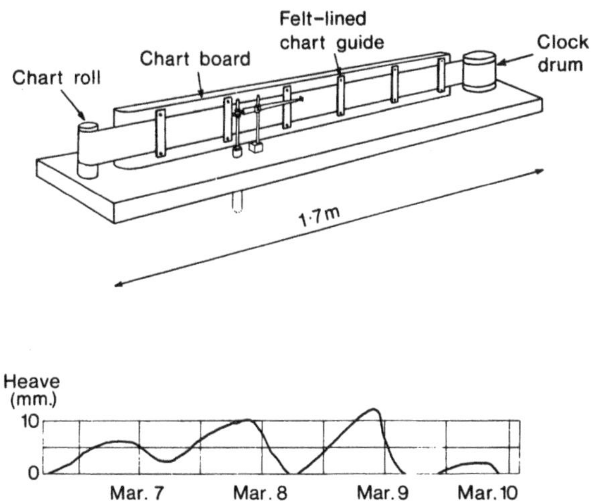

3.6 Example of a clockwork mechanism for use in obtaining a continuous trace of frost heave in soils (after James 1971).

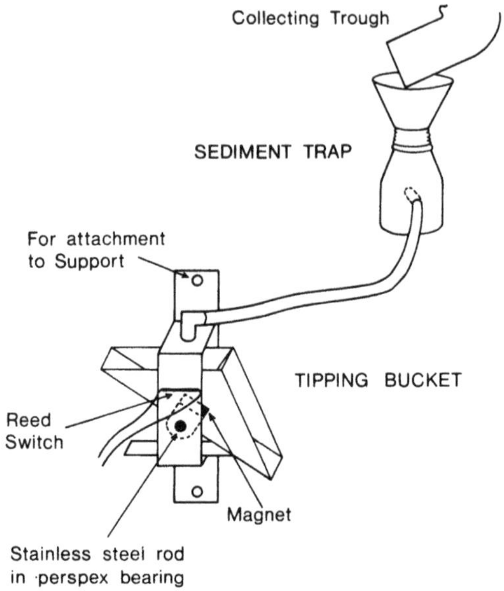

3.7 The tipping bucket mechanism to record event occurrence, in this case for throughflow measurement (after Knapp 1973). Signals sent by the reed switch are recorded on a data logging system which includes the timing of events.

Recurrence

Continuous or regular discrete observations are still relatively uncommon in geomorphology because of the heavy demands they make on cost and labour. On the other hand reports of past events, their magnitudes and period of operation are generally more available. From this data and from the spacing of events from continuously recorded data a body of information called recurrence time is available. Here time is the variable of interest and is measured between occurrences of the event.

An example of this, which has been widely used in fluvial mechanics, is the idea of rest time, the time between successive jumps of a particle lying on a stream bed. Information of this type can be characterized by the distribution of recurrence times and certain common distribution patterns can be described by relatively few parameters (the moments of the distribution, p. 72). The greatest use of this information has so far been made in hydrology, mainly because more data is available. Here elements of continuous series may be treated as discrete events by considering particular, presumably meaningful, thresholds. Thus the distribution of time between which particular discharge events occur or are exceeded is frequently of interest and is usually expressed as the percentage of time than an event magnitude is equalled or exceeded (fig. 1.7). Hydrological studies of drought events (when flow ceases) can include a consideration of zero events. In these cases the continuous pattern and the situation are more similar to recurrence times in mass movement. Recently, techniques have become available to establish recurrence intervals for data which include zero flow situations (Jennings and Benson 1969). Surges occurring in glaciers are comparable phenomena to storm hydrographs in rivers when considered with respect to their occurrence in time and can be treated in analysis as recurrence times.

It will be readily apparent that in long-term chronology there is now available a useful armoury for attacking the problems of position in time and of the passage of time. There is, even from those studies related strongly to hydrology, very little information on variables operating in time, on recurrence times and on times between events. This reflects in part the use of 'incidental' information; in part the paucity of thought given to sampling in time and experimental design generally, in part the heavy cost in time and money involved in continuous recording and, perhaps most important, in part in the relatively late development of process studies and precise collecting and recording of information in geomorphology compared with other sciences.

The more information becomes available, the more easily will geomorphologists be able to design experiments which characterize adequately the events which they wish to investigate. Emphasis so far has been on the analysis of data collected in time rather than the sampling framework appropriate for their collection. One must supplement the other if predictive capabilities are to improve. In hydrology great advances have been made in the analysis of regularly collected data by national authorities but even here there are features of interest, particularly the storm hydrograph, for which experimental design situations are inadequate on other than continuously monitored stream systems (Wolf 1966).

Records of runoff, sediment yield, precipitation, snowfall, water quality and many other variables are now being collected all over the world for management or research purposes. It is becoming important, for economic reasons, to optimize the data collection procedures and it should not be assumed that expensive, continuously monitoring systems are necessarily the most efficient methods available. The central question is the possibility of designing temporal sampling schemes so that *discrete* data, collected with less expensive equipment and for a shorter time and at less cost, can provide information to a required efficiency level. The object is to collect data which yield all the necessary characteristics of the system but avoid the problems of insufficient data (incomplete record), of redundant data (add nothing to the record) or of persistence (carry-over) effects in the record. Added benefits are also felt in analysis where calculation time may be considerably eased if the same answers can be achieved with less data. Redundant data and consequent analysis cost form a common problem where processes are monitored continuously.

The design of such discrete data collection systems is of course primarily dependent on the research questions. For example, information on the detailed behaviour of a flood might require short (perhaps half-hourly) discrete observations, but long-term budgets of glacier behaviour might be obtained using daily, weekly or longer records. Several techniques are now available to facilitate decisions on the most efficient sampling interval for a particular job. The most commonly-used methods are based on a consideration of the information content of time-variant data (Matalas 1969, Quimpo and Yang 1970, Gupta 1973). Conceptually, the information content of a statistic is inversely proportional to the variance of the same statistic. The technique helps us to choose the most suitable sample interval by discovering the time interval between sampling which will yield minimum variances of the mean or the sample variance and thus maximum information content.

The properties of the sampling distributions are given by:

$$V(\bar{X}) = \frac{\sigma^2}{N} \tag{3.1}$$

$$V(S^2) = \frac{2\sigma^4}{N} \tag{3.2}$$

and the information statistics for the variance of the mean (I_1) and the variance of the sample variance (I_2) by:

$$I_1 = \frac{V(\bar{X})}{V^1(\bar{X})} \tag{3.3}$$

$$I_2 = \frac{V(S^2)}{V^1(S^2)} \tag{3.4}$$

where V = variance
N = sample size
σ^2 = population variance
$\bar{X}$ = sample mean
S^2 = sample variance

and $V^1(\bar{X})$ or $V^1(S^2)$ are the variance of the mean and the sample variance according to the appropriate persistence model (e.g. First Order Markov model). Details for calculation of these models are provided by Gupta (1973).

The information content will change with changes in sampling interval and a consideration of the curve plotted for such changes (fig. 3.8) enables a decision on optimum sampling frequency to be made.

In practice the sample interval will be chosen on economic and logistic grounds determined by the information content expected. In such cases we will be able to state that our logistic decision will result in a certain information content. It is, however, also possible to choose a pre-set (say $I_2 = 0.8$) information content requirement and thereby determine the interval between recording visits in the field. More frequent observation leads to redundant data, less frequent to loss of data, at the chosen information threshold.

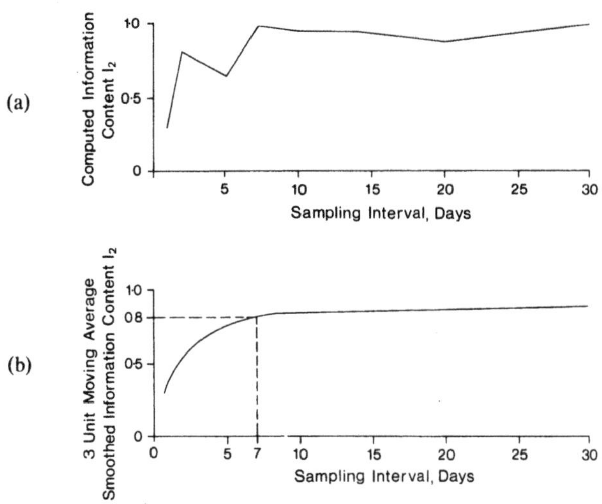

3.8 (a) Variation in computed information content with sampling interval. (b) Variation in smoothed information content with sampling interval, e.g. a preset value of $I_2 = 0.8$ yields a sampling interval of 7 days (from data for the Eel river by Gupta 1973).

The technique is at the moment applied in water resource and management networks but there seems no reason why it should not be used in geomorphological studies of other processes. Indeed, for any long-term study with high demands on cost or labour or involving harsh field conditions and difficult access it seems sensible to design pilot studies whose express purpose is to collect data suitable for the design of the future permanent network.

4 Analysis of temporal data

Introduction

Geomorphological data are of two types, serially independent and serially dependent. In serially independent data the values at one time are not affected by those occurring before or after that time for a given time scale and the observations may be regarded as discrete events. For example, the total volume of snow involved in an avalanche at a particular place is likely to be completely independent of that at the same place several years later. Similarly, the volume of water yielded in a particular flood event might, because of a random component in the precipitation and hence in the antecedent soil moisture conditions, be completely independent of the water yield in another flood event. Data of this type are amenable to the common methods of frequency distribution by the methods of moments.

Most phenomena, however, demonstrate a certain amount of dependence on the magnitude of preceding events. The question of dependence and independence is a matter of careful definition of the variables under consideration and the time scale of the observation, a point already made in reference to the work of Schumm and Lichty (p. 13). Consider the discharge of a stream. On the scale of a few hours in the passage of a flood (fig. 1.2, p. 6) each observed value of discharge appears related to the events immediately preceding and succeeding it. In the longer time period, say a week or a month, the high discharge events are separated by long periods of base-flow and hence appear independent of each other. On this scale total discharge of storm 1 is considered independent of that in storm 2. For an even longer period of time, for the first six months of the year in this case, there is an overall recession of the base-flow and the individual 'instantaneous' hydrographs make a relatively small contribution to the monthly flow. Each flow again appears related to the earlier flows. Where the data are related to one another the assumptions involved in normal statistical analysis are invalidated and recourse has to be made to a special set of techniques known as time series analysis, which takes into account this serial dependence.

For both non-serially and serially dependent data, where the mean and variance remain constant, the series is said to be stationary. Variations in mean and variance are easily observed by plotting the raw data, though formal tests are statistical. Krumbein (1966) likens this question of dependence and independence to memory. He sees a sort of sliding scale with absolute independence at one end, with no built-in effect of previous events and values, to absolute dependence at the other where all the events are determined and utterly dependent. In the latter case the system is thought of as having a very, almost infinitely, long memory. Variations in dependence or independence are not so easily observed and numerical methods are available for their differentiation.

Conventional statistical analysis compares the degree of association between two independently distributed random variables by means of the correlation coefficient. For example, in the plot of valley side slope and channel slope (fig. 4.1) observed at different locations in a valley system the strength of the statistical relationship may be evaluated by the correlation coefficient. The latter may range between −1 and +1 through 0. When it is close to −1 and +1 the variables are said to be highly correlated; conversely as the value approaches zero, correlation is said to be very poor. In the example (fig. 4.1) the correlation coefficient was found to be 0.84 which is highly significant.

Imagine now, that instead of different variables we observe the *same* variable through time (table 4.1). Thus we might be observing the amount of melt at a glacier snout every day. A set of 'paired' observations could be

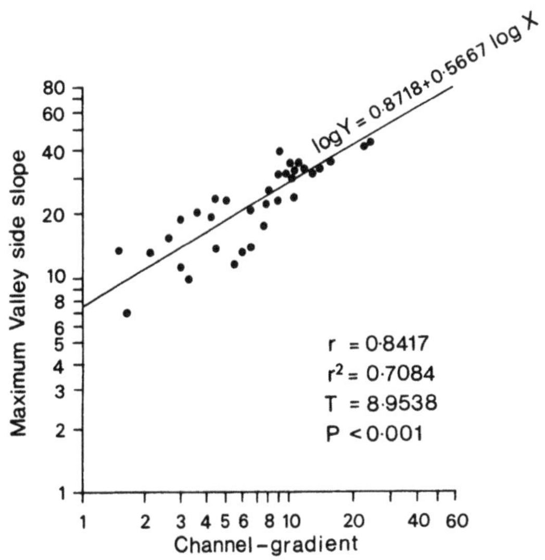

4.1 Log-log regression of independently observed random variates (maximum valley side slope and channel gradient) (after Thornes, 1967).

Table 4.1 The amount of daily melt at a temperate glacier snout in cms; meltwater equivalent shown as pairs of lagged observations.

Raw data, melt in water equivalents (cms)	*For lag 1 compare:*		*For lag 2 compare:*		*For lag 3 compare:*		*etc*
15	15 and	10	15 and	10	15 and	15	etc
10	10	10	10	15	10	10	
10	10	15	10	10	10	11	
15	15	10	15	11	15	6	
10	10	11	10	6	10	3	
11	11	6	11	3	11	7	
6	6	3	6	7	6	4	
3	3	7	3	4	3	etc	
7	7	4	7	etc			
4	4	etc					

obtained by observing the values for days 1 and 2, 2 and 3, 3 and 4, 4 and 5 and so on. These observations are said to lag by one or to be the lag 1 set. The lag 2 set may be obtained by taking 1 and 3, 2 and 4, 3 and 5, 4 and 6; and lag 3 by comparing 1 and 4, 2 and 5, 3 and 6 and so on. For each set of paired observations the correlation coefficient may be found. It is now called the autocorrelation coefficient since it is the correlation of one variable with itself over lagged intervals. The relationship of the autocorrelation coefficient against lag is known as the autocorrelation function or acf. If the elements of the series are completely independent the values of the coefficients will not be significantly different from zero. The exception to this is at lag 0; here each observation is correlated with itself and therefore the coefficient must have a value of one. Where serial dependence in one form or another does exist it can usually be recognized in the correlogram (the plot of the acf).

Discrete, serially independent events

A set of observations may be described in terms of the number of observations falling into each of a number of classes. The data are plotted on a histogram usually with each class represented by columns of equal width but with their height proportional to the number of observations in that class. If the numbers in each class are added successively the result is the cumulative distribution function (fig. 4.2a and b). If the number per class for each class is divided by the total number of observations, then we have the probability of obtaining a value for any one class. Thus in the example where we have obtained the histogram from a given large number of particle long-axis measurements figure 4.2c represents the probability of obtaining a particle in any size class when the large number is sampled at random. This is the probability density function (pdf). If the probabilities are summed cumulatively the cumulative probability distribution (fig. 4.2d) is obtained. This yields (on the y-axis) the probability of getting a value equal to or less

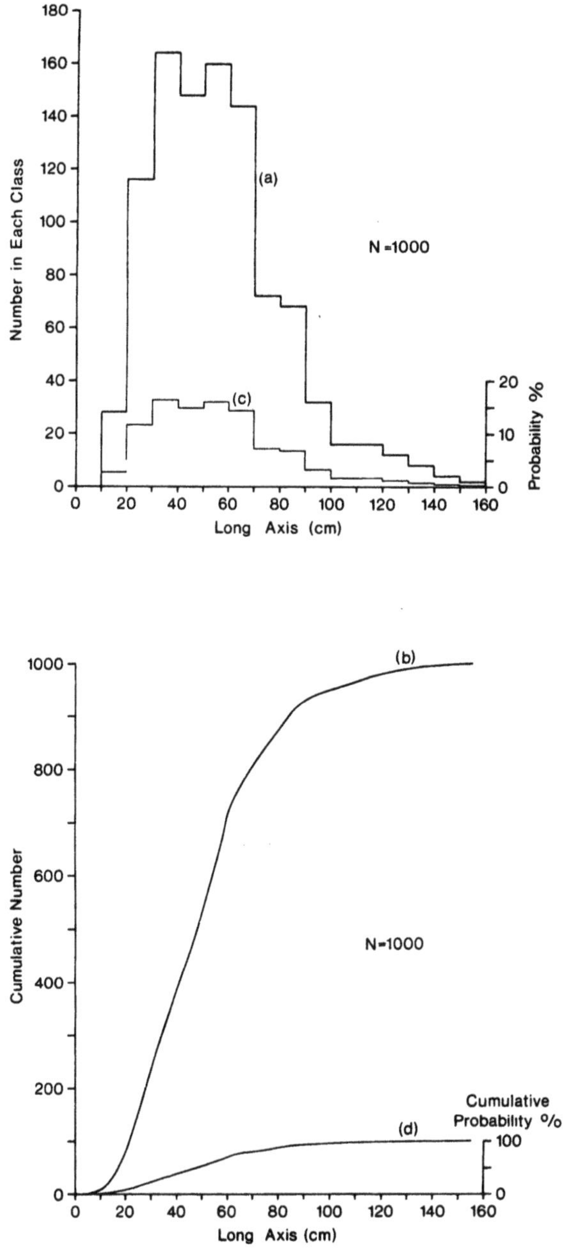

4.2 Methods of depicting frequency count information. (a) Frequency histogram. (b) Cumulative distribution function. (c) Probability density function. (d) Cumulative probability distribution.

than the value specified on the x-axis. The probability frequency distribution is important for describing events in time. The important point, however, is that the probability density functions of most variables can be described by a few basic types of function. If the basic type is recognized, then the pdf can be described by one or two simple values called the parameters of the distribution.

Certain kinds of 'process' are known to generate the basic types. Thus, for example, if we shake a dice and record whether or not a six resulted, the process is said to be a Bernoulli type process. In recognizing the basic type to which our distribution has strong affinities, therefore, we not only seek for a few simple parameters to describe it; in addition it is possible that a process analogous to the known statistical generating process may be found.

An an example, certain river run-off data are said to exhibit a Markov process. It is suggested for example that in hydrology this property in daily discharge data represents the 'carry-over' in channel and soil moisture storage from the values observed on the previous day.

We have perhaps laboured this point but it is important to remember that probability distributions are used for interpretation, forecasting and model building as well as describing the data of interest.

In geomorphology we have interest in two further kinds of data. The first kind records the occurrence and non-occurrence of an event, the number of events per unit time and the time between events. In the second type attention turns not only to the frequency of the events but also to their magnitude. The simplest type of question we can ask of the first type ishow many times (k) out of a given number of times (n) shall we find the system in one or other of two conditions. In situations where there are only two possible outcomes, e.g. head or tail in the toss of a coin, the process is said to be a Bernoulli process. For such a process each event is to be independent, there must be no 'pattern' in the succession of events and the long run probability of one condition (p) or another condition (q) must be known. Consider as an example a solution hollow observed once a week throughout the year in a non-seasonal environment. On each visit the hollow has or has not water in it. It is possible to say, that the overall probability of finding water in k times out of n visits (e.g. 5 out of 25) is given by the basic Binomial distribution. This is expressed by:

$$p(k) = p^k(1 - p)^{n-k} \frac{n!}{k!(n - k)} \tag{4.1}$$

$p(k)$ is the probability of water being in the hollow k out of n times and ! is the factorial sign. As another example imagine we have an experimental plot for rainfall erosion. If a particular square centimetre receives a raindrop in one second out of every ten on average, we can evaluate the probability that it will receive a drop in five one-second intervals out of ten, or twenty one-second intervals.

As the probability of the event becomes relatively small and the period between events large, the distribution is better described by the Poisson distribution. Similarly, events which are rare in space could be described by the Poisson. For example, the probability of finding a waterfall in a stream divided up into a large number of unit lengths is likely to be such a case. In the temporal sense the occurrence or non-occurrence of rapid failure of a clay slope per unit of time on a section of cliffed coast and the distribution of probabilities of a glacier surge ought both to be described by the Poisson. Data which records if a particular threshold level in a system is or is not exceeded (without reference to the *magnitude* of the level attained) can be characterized easily by modified Poisson models. The general form of the distribution is given by:

$$p(k) = e^{-\lambda}(\lambda)^k/k! \qquad (4.2)$$

where λ is the mean rate of occurrence, k, the number of observations and e the base of natural logarithms. In figure 4.3a we show the distribution giving the probability that a river bluff will be undercut 0, 1, 2, 3, 4, 5 and 6 days in a year given that it is normally undercut on 2 days a year. Obviously the probability of being undercut on a relatively large number of days decreases quite rapidly. Figure 4.3b gives the probabilities of undercutting if the average number of days is 4. It will be seen that in this case the distribution begins to look more normal because the event is becoming more common.

Another way of looking at the data is to ask the question: what is the time distribution *between* events? Thornes (1971) considered scree slopes to be analogous to queues; in this model the rate of arrival and departure of

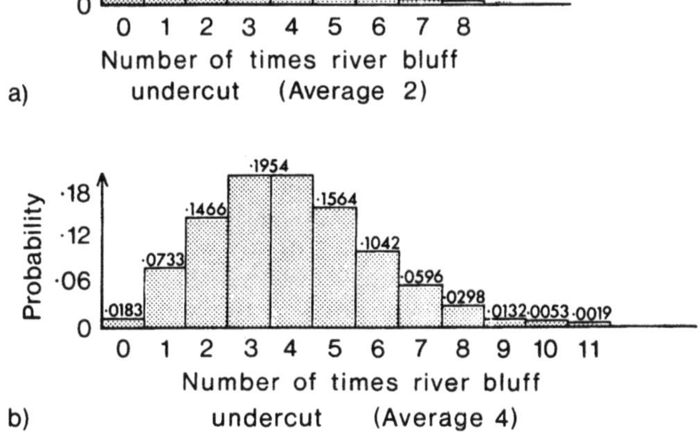

4.3 (a and b) Effects of changing average value on the Poisson distribution.

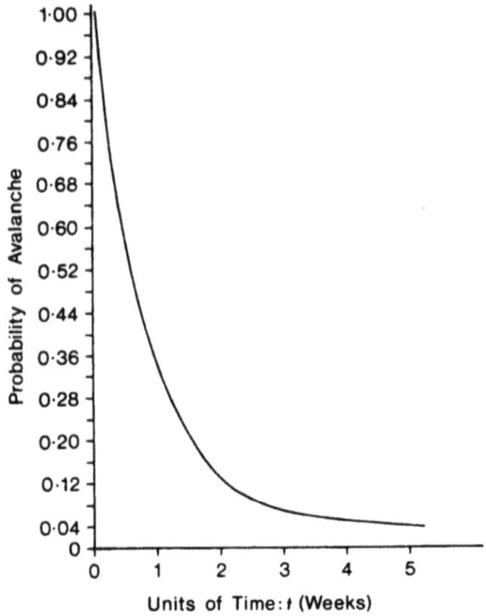

4.4 Cumulative exponential distribution for the probability that time between arrivals of avalanches exceeds *t* units of time when mean arrival rate is one per week.

material becomes important for determining the 'length' of the queue. Given an average waiting time between arrivals λ we expect the distribution of these waiting times to be related to the Poisson and in fact it is; through the exponential frequency distribution (not density function) which has the form

$$F(t) = e^{-\lambda t} \tag{4.3}$$

where $F(t)$ is the probability that the length of time between successive arrivals will be equal to or greater than t. This is a very useful graph since it can be used to state the probability that the time between successive arrivals will be of a given length. In the case shown (fig. 4.4) λ is given as 1 (1 avalanche per week). We can read from this hypothetical graph the probability of 3, 5 etc weeks between the successive avalanches.

Two further distributions are of particular interest in geomorphological work on events whose magnitude are not specified. The first, the negative Binomial, answers the question, how long, or how many units of observation (trials) will be needed to record a single event? The second, the Gamma, is useful in yielding the distribution of times up to the *r*th event. Suppose for example an experimental design requires that 25 observations of a particular event are required and that the mean rate is λ (e.g. 5 per week), then it will be of interest to know the probability of having to wait, 5, 10, 14, 15 etc weeks before the required number are observed. This might be especially relevant for logistically difficult or economically expensive investigations such as Arctic run-off.

If we attach to each observation a value then the distributions are twofold; there is a frequency distribution and a magnitude distribution. The product of frequency and magnitude is work. Sometimes the frequency will be relating to independent events, whereas the magnitudes vary regularly, for example where there are seasonal variations in values of climatically influenced processes. In other cases while the magnitudes are independent, the events themselves may be clustered in time. Rock fall events may vary randomly and independently in magnitude but be clustered in time. In addition to this complication magnitudes are almost invariably continuous whereas events may be quite discrete. Magnitude data may also be treated in the discrete form especially where significant threshold or capacity values exist. In soil moisture for example a threshold exists when infiltration capacity is reached. In most situations, however, the essential continuity of the possible magnitudes leads at least to persistence and often to a high degree of serial dependence. Thus even quite a large 'event' will have associated with it related 'falling-stage' values. This is partly a function of scale as we mentioned earlier with reference to the hydrograph. Partly also it is related to the inertia in a system which may result in relatively long relaxation times.

It is thus for the truly extreme events that magnitude and frequency analysis, via the distribution and density functions, may be most usefully employed. There is a basic problem that extreme events are not only large but also rare, so that we have no idea of the basic distributions to which they belong. Ironically also, the longer the period of observation the more likely it is that the basic statistical requirement of stationarity is lost. Gumbel (1958) attempted to by-pass the initial distributions in order to reach a description of the extreme values. His work has received widespread prominence amongst hydrologists and meteorologists partly because of the versatility of the three basic extreme value distributions but also because of the facility of fitting data to special probability paper. It is also rather teasing because the versatility of the Gumbel distribution essentially suggests its relative insensitivity.

The notion of frequency and magnitude is particularly important for comparative work. It is generally held, for example, that discharge events of different frequency (and hence recurrence interval) are responsible for different aspects of channel geometry. Similarly, differing recurrence intervals for events of the same magnitude form a basis for comparative study. As Hewitt (1970) has pointed out there has, as yet, been little development of the idea applied in a spatial context. An extreme event of low recurrence interval in one area may be a common event in another environment and vice-versa.

The recurrence interval is hypothetically the average time between events of the same magnitude. A 50-year recurrence interval implies that the event is only likely to occur once in 50 years. While this is essentially a probabilistic notion, the plot of recurrence interval against magnitude (fig. 4.5) is essentially based on a sample of one observation for each magnitude reported.

The raw data is usually for annual or less frequent events, the most common being the annual flood series. This consists of the highest discharge

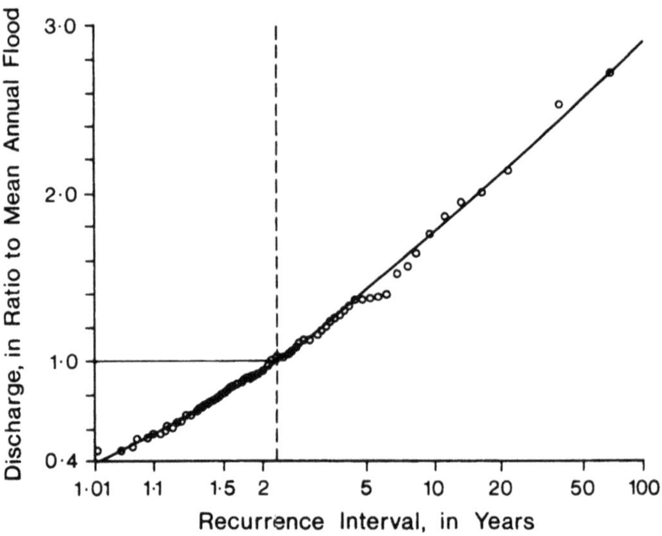

4.5 Recurrence interval in years for the Youghiogheny and Kiskiminetas rivers (after Leopold, Wolman and Miller 1964). It will be seen that the mean annual flood has a recurrence interval of 2.33 years.

for some fixed time period (e.g. 24 hours) which occurs in each year of the records. In 60 years of records there will be 60 observations. In a partial duration series on the other hand all the discharges above a particular level may be used. Obviously, there are many values of extreme events which can be used. Workers must take care to ensure, as far as possible, that they are not dealing with mixed populations whose characterizing distributions may be quite different. For example, run-off in Mediterranean areas arises from spring snow melt as well as from intense summer rains and the distribution properties for each ought to differ.

The events are normally arranged in order of decreasing magnitude, the largest being rank 1 (see table 4.2), the second largest rank 2 and so on. From this table the recurrence time T and the probability $P(\mathrm{x} \geqslant \mathrm{X}_m)$ that a magnitude equal to or greater than a particular value is obtained are given by

$$\mathrm{F(x)} = P(\mathrm{x} \geqslant \mathrm{X}_m) = \frac{m}{N+1} \quad \text{and} \quad T = \frac{N+1}{m} \tag{4.4}$$

where m is the rank (e.g. 1st, 2nd, mth) in the table when events are listed in order of magnitude; N the total number of observations. For example, if a precipitation of 11 mm in 30 minutes is 3rd in a series of annual maximum 30-minute precipitation values from a 24-year record, its recurrence interval will be 8.33 years and the probability of this value being equalled or exceeded is 0.12. This may also be expressed by saying that a 30-minute precipitation of 11 mm is likely to be equalled or exceeded not more than 12% of the time.

Table 4.2 Maximum annual rainfall values 1940-61 for Madrid, rearranged for ranked order and including other data necessary to demonstrate the basis for a partial duration series (Madrid 1940-61) (data from Elias 1963)

Maximum annual rainfall values for 30 min. periods for 1940-61 Madrid in mm			*Ranked values of maximum annual rainfall 1940-61 in mm*		*Partial duration series. All data above 8.7 mm in 30 mm. periods (1940-61)*	
Value	*Month/Year*		*Value*	*Month*	*Value*	*Month*
16.7	O	1961	25.0	Jl	25.0	Jl
12.1	O	1960	21.9	Ag	21.9	Ag
17.3	J	1959	17.3	J	18.5	Ag
10.7	D	1958	16.7	O	17.3	J
4.5	O	1957	16.6	S	16.8	O
7.2	S	1956	15.6	Jl	16.7	O
21.9	Ag	1955	12.8	My	15.6	Jl
12.8	My	1954	12.3	D	15.4	S
6.8	J	1953	12.1	O	14.4	Ag
6.0	N	1952	11.1	Jl	12.8	My
9.3	J	1951	10.7	D	12.5	Ag
15.6	Jl	1950	9.7	S	12.1	My
9.7	S	1949	9.3	J	11.2	Jl
8.3	F	1948	8.3	F	10.7	D
25.0	Jl	1947	8.1	A	10.4	Ag
5.3	My	1946	7.9	S	10.0	S
12.3	D	1945	7.2	S	9.7	S
16.6	S	1944	6.8	J	9.7	S
8.1	A	1943	6.5	N	9.3	J
7.9	S	1942	6.0	N	9.0	My
6.5	N	1941	5.3	My	8.8	J
11.1	Jl	1940	4.5	O	8.7	My

The data in columns 2 and 3 on table 4.2 may be plotted on graph paper with magnitude on the ordinate (y-axis) and recurrence interval on the abscissa. Special graph paper may be obtained which is drawn up in such a way that data conforming to a particular distribution will plot as a straight line. Two examples of this are log-probability graph paper and Gumbel extremal probability paper. In the former, data conforming to the distribution known as log-normal, plots as a straight line; some discharge data of greater frequency than extreme events have this property. In the latter, data having a Gumbel (type 1) distribution expressed by

$$p = \exp(\mathrm{e}^{-\mathrm{e}^{-y}}) \qquad (4.5)$$

plot as a straight line. An example of this is given in figure 4.6 where the annual series for 30-minute precipitation maxima at Madrid (1940-61) have

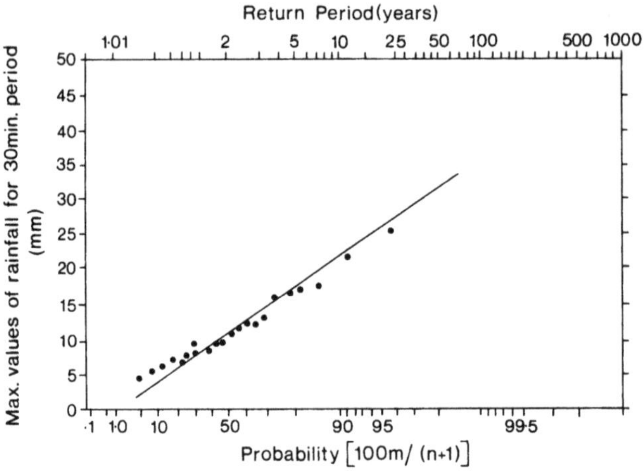

4.6 Gumbel distribution of annual maximum 30 min precipitation partial series.

been plotted. These conform tolerably well to the Gumbel distribution and from the graph various parameters of the distribution may be estimated. Alternatively the values for various recurrence intervals may be identified by obtaining the value of y using nomograms (Weiss 1955). Another commonly used extremal distribution is the Pearson type 3.

These basic forms differ essentially in terms of the initial distribution which is assumed. In the application of this type of fitting procedure there are two main sources of error, the inadequacy of fit to the theoretical distribution and the errors arising from 'sampling'. As a general rule the observed curve should not be extrapolated beyond a return period greater than twice the period of observation.

Serially dependent information

There are three basic types of serial dependence, trend, periodicity and autoregressive moving average dependencies or persistence.

Trend

Where trend exists it is often easily recognized by a simple plot of values of the series against time. For example, the annual mean values of nitrate concentrations for the river Stour in Essex for 32 years show that, overall, the concentrations have increased fairly regularly through time (fig. 4.7a). The trend is linear and positive (nitrate is increasing regularly as time passes). At the same time there is some 'scatter' about the line indicating variations about the trend. It is clear that the data are non-stationary. Trends need not be linear. The plot of mean sea level (Mörner 1971, fig. 4.7b) for the last 9000 years shows a clear non-linear relationship in that the rate of increase of sea

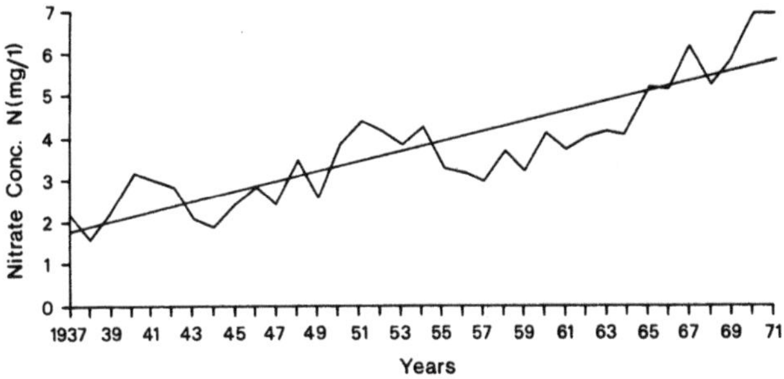

4.7(a) Fluctuations in mean annual nitrate concentrations for the river Stour 1937–71.

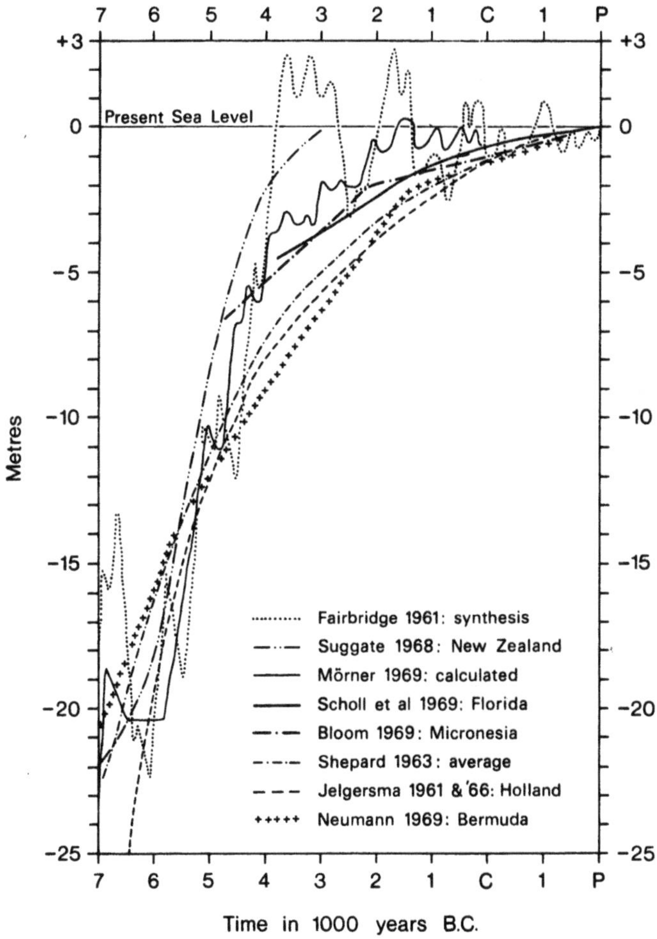

4.7(b) Sea level change in the last 9,000 years according to various authors (after Mörner 1971).

level per unit of time decreases with the passage of time. Time trends, both linear and non-linear are well known from the geomorphological literature. The scatter about the line partly comprises imprecise dating, partly it arises from the complexity of forces acting to produce output of the system (the observed values of the variable) and partly from poor measurement of the variable in question.

The trends can be described by the procedure of regression by least-squares. This involves calculating the equation of the line which best fits all the points in the plot. The criterion for best-fit is that the vertical distances from each point to the line should be made a minimum when they are added together. In other words the line is 'moved about' until the summed distances cannot be made any smaller. The linear regression then has the form $Y_t = kT + C$, where Y_t is the value of the variable in year t, T is the elapsed time since the beginning of the record, k the slope of the best-fit line and C the value of the variable at the beginning. Thus for nitrate concentration on the Stour, the equation is $N = 0.11576T - 2.51072$, where N = nitrate concentration, T = time in years after 1900 (record starts in 1937) (fig. 4.7a). An example of a non-linear time trend is given by the equation $Up = Ce^{-kt}$ where Up = uplift since deglaciation, t = time expressed in 10^3 years since deglaciation, k and C are constants; this was found by Andrews (1970) to express the rate of uplift for a particular point in the Canadian Arctic.

Another type of non-stationarity occurs when a system undergoes some radical change in its nature or rate of operation. For example, suppose there is a regular fluctuation in sediment yield for a small basin according to the seasons. The mean and variance in sediment yield observed, say on a monthly basis, may remain stationary for several decades. If rapid and extensive changes then occurred in the character of the vegetation cover, the sediment yield could change dramatically both in average value and in the extent of seasonal fluctuations. This is shown hypothetically in figure 4.8. Behaviour after the effects of the vegetation removal are felt, in this case could be

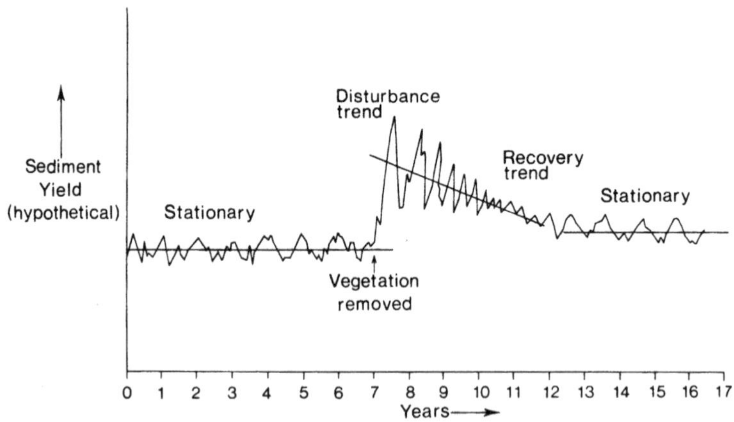

4.8 Hypothetical response of sediment yield to devegetation.

described as damped oscillatory behaviour. Eventually the system, as shown, reaches a new fairly stable and statistically stationary condition. Such changes in systems leading to non-stationarity are common in series with thresholds.

Periodicity

Many geomorphic processes are controlled by regular climatic phenomena and exhibit cyclic or regularly periodic fluctuations. Sediment supply, discharge, soil erosion, solution, soil creep are but a few examples. Other regularity occurs from the existence of thresholds of stress. Gradual build-up to some stress level followed by failure and fresh build-up, as appears to occur on steep debris with a constant supply, could exhibit this type of regularity. Unfortunately, geomorphological (as opposed to hydrological) data indicating this type of phenomenon are relatively rare. One example is a record for an Antrim mudflow (Prior and Stevens 1972) (fig. 1.12). There appear to be two reasons for this, the logistic difficulties and the mechanical problems involved in continuous monitoring. Most continuous data available to, and used by geomorphologists is of run-off and sediment yield. It is no accident that some of the most useful models and results are from this area of research.

Where regularly periodic phenomena exist they can frequently be described by the relatively straightforward technique known as Fourier analysis. Consider for example the discharge of the river Stour at Langham (fig. 4.9a). It will be noticed that for the six-year record there is a regular high

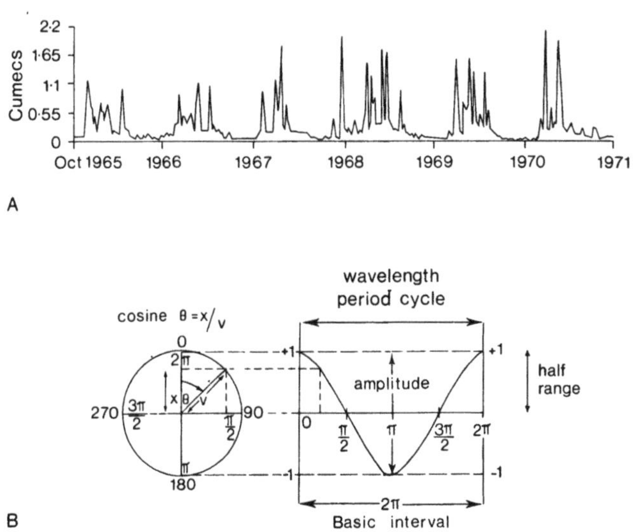

4.9 (a) Discharge of the river Stour at Langham, Essex. (b) Notation for the Cosine curve.

discharge in the winter part of the year and a corresponding low in summer. Fourier analysis seeks to describe such data as closely as possible by trigonometrical functions based on the sines and cosines of elementary trigonometry. The cosine curve (fig. 4.9b) is obtained by rotating a radius through 360° and obtaining the cosine of the angle between the radius and the vertical. This is positive in the first and last quadrants and negative in the second and third. This complete rotation gives one cosine curve and is equal to 2π radians (360°), the basic interval. By using the formula $\cos 2\theta$ where θ is the angle two complete curves can be fitted into the basic time; $\cos 3\theta$, three complete curves and so on. The number of waves per basic interval is known as the frequency. For a frequency of two, it takes $\frac{1}{2}$ the basic interval for one wavelength to be completed; for a frequency of three, one wavelength will be completed in $\frac{1}{3}$ of the basic interval. In other words the period for one wavelength is the reciprocal of frequency. To match the data set we need to multiply the amplitude by a constant A, so we have the expression $Y = A \cos \theta$. Finally, because the curve need not start with its maximum at zero, we have to allow for it to be shifted to the right by substracting an amount θ (phi); this is an angular measure which varies according to whether we are fitting 1, 2, 3, 4 etc complete curves in the basic interval (i.e. it varies according to the frequency).

Consider again now the data for the Stour at Langham (fig. 4.9A): observations on discharge were obtained every seventh day so there are 312 data points for 6 years. Because there is a clear annual cycle, we might expect 6 oscillations in the 6-year period and indeed we find that using a best-fitting technique on the data, a model which allows for 6 oscillations gives the best fit. The actual equation is expressed in cosine and sine terms as a result of algebraic manipulation as follows:

$$Y = C + a[k]\cos(k\theta) + b[k]\sin(k\theta)$$

$$\text{where} \quad a[k] = A[k]\cos(\Phi[k]) \qquad (4.6)$$

$$b[k] = A[k]\sin(\Phi[k])$$

Y is the value at a distance 0 along the basic interval
k is the number of oscillations in the basic period
C is the mean of all the observations
$\Phi[k]$ is the phase angle of the kth harmonic
For the 6th harmonic (6 cycles in the 6-year period) the equation for the Stour data is

$$Y = 67.834 - 36.26\cos(6\theta) + 48.45\sin(6\theta) \qquad (4.7)$$

Obviously this technique has fruitful application where there are very regular cycles in the data. In geomorphology this is often the case in discharge, mean annual sediment yield, temperature-controlled phenomena and many types of coastal data. Often instead of fitting a single harmonic, such as the 6th, we sum together several harmonics. This enables more complicated curves to be fitted to the original data.

The situation sometimes arises in which the basic regularities cannot be determined a priori; they may not even be very obvious from an initial inspection of the raw data. In this case, the procedure of fitting successive harmonics up to fairly high frequencies may clarify the issue. The harmonic or combination of harmonics which best fits will minimize the summed distances between the actual data points and the fitted curve. Another way of expressing this is to say that the variation in the raw data is most adequately accounted for by the line which fits the raw data points best. If for every frequency which is fitted we plot the amount of variation accounted for, then the 'best-fit' curve will have the highest peak on the plot. Such a plot is called a periodogram and was developed by Schuster. This technique, of isolating the separate harmonics present in the data and showing how much variance each frequency accounts for, is classical Fourier analysis. Each harmonic is considered separately and discretely. Since we must have at least 2 data points the highest harmonic which can be defined is $N/2$ where N is the total data length.

Instead of identifying only the principal harmonics, such as the 6th (annual) harmonic in the Stour data, we should like to range across all possible frequencies and so have a continuous plot of the variation explained by any and all frequencies up to that defined by $N/2$, in this case the 156th harmonic ($N = 312$ data points for 6 years). In our example this is a period of 2 weeks or twice the basic unit of observation (7 days). A plot describing the variance accounted for by all frequencies is called the variance (or power) spectrum and the technique for obtaining it, spectral analysis. The power spectrum for the Stour data is shown in fig. 4.10. Although we have presented it as the logical extension of harmonic analysis, it can be reached more directly by a procedure known as Fourier transformation.

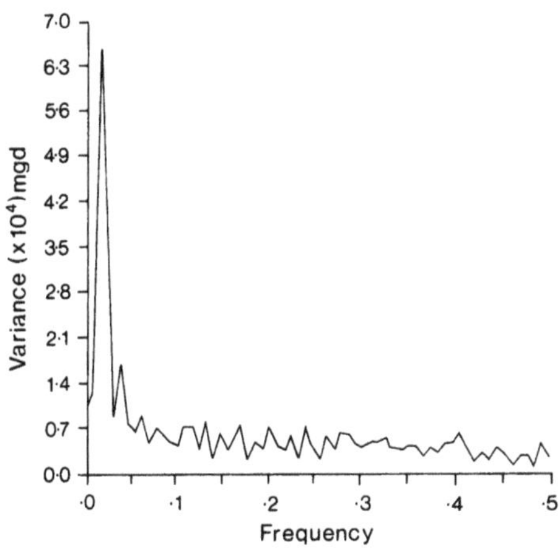

4.10 Variance spectrum for the discharge of the river Stour at Langham, showing the high variance of the annual cycle.

Although the Fourier transform is an equation with numbers, it is best regarded at this level as an operator, i.e. it performs an operation. Examples of operators are +, — and ÷. The operation it performs is to change time (period) based information into frequency information. Recall that the correlogram was a plot of the autocorrelation coefficient against the lags between observations. One lag could equal one week, so that twelve lags equal three months. This is time based. The autocorrelogram is based on the auto-covariance function which describes the amount of variation for each lag. It will be no surprise therefore to learn that the variance spectrum (variation *vs* frequency) can be obtained as the Fourier transform of the autocovariance function (variation *vs* lag). The transformation is also reversible, the autocovariance function can be obtained from the spectrum. This useful property of the Fourier operator is used time and time again for manipulating time series data.

Persistence

We have described so far long-term trends and regular periodicities and some of the techniques for handling them in a series of observations on one variable through time, such as movement at the snout of a glacier. There is a third type of dependence in data which is called persistence. This component is produced in one of two ways, either by the presence of some physical factors in the system which produces the output series; or as a result of the way in which we collect the data and subsequently manipulate it.

One of the easiest examples of physical persistence to understand is that produced in the flood hydrograph. With a plot of discharge against time, surface run-off events may be separated by relatively long periods of uniform base flow. The channel parameters, especially depth where banks are cohesive, remain fairly constant. Assume (fig. 4.11a) that with a large storm there is rapid bed scour, this is a response to the changed discharge. It could be that the readjustment is relatively slow, so that even though the storm has passed the channel takes several days to reach its former condition of depth. In other words, the effects of rapid scouring persist for several days after the event. The first day after, the effect would be fairly high and it would gradually diminish until, say, after the sixth day its effects have completely disappeared. With a large number of comparable events in a channel, the theoretical autocorrelation would appear as in fig. 4.11b. It has a high coefficient for lag 1 which decreases very rapidly until by lag 6 it is no longer significant. Persistence effects of this type are quite common in geomorphology and some were described in the discussion of relaxation time and paths in chapter 1 (p. 16). Where this short term persistence is present the series is said to exhibit autoregressive properties. Different types of technique are required for investigating long term persistence and these have been very slow to develop, though Mandelbroot and Wallis (1968) have made special efforts to analyse them.

It is not too difficult to appreciate that in collecting and analysing the data we may consciously or unconsciously affect the autocorrelogram and

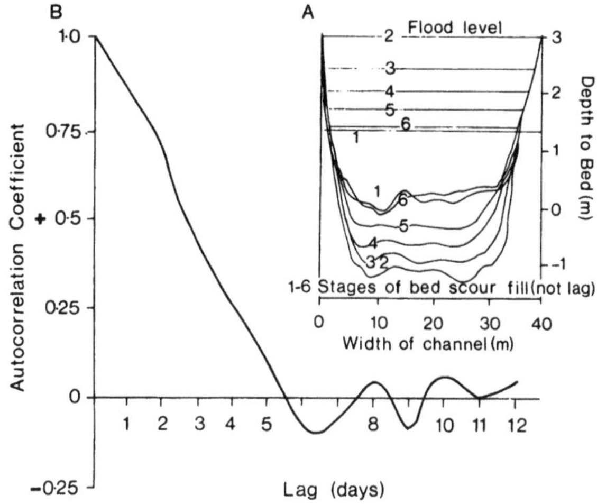

4.11 (a and b) Hypothetical correlogram for height of channel bed. (a) After Leopold *et al.* 1964.

through it the power spectrum. For example, if we have a daily discharge series for say 100 weeks and we average the data in some way, we will make adjacent values in the new series relate more closely to each other. The types of averaging are shown in figure 4.12a. In cascaded averaging, we average the averages from the previous step. In running averages we 'overlap' observations and hence smooth the resulting values. In scale averaging we use successively larger separations between observations. This operation tends to induce autocorrelation at higher and higher lags according to how much we average; the effects on the spectrum are shown in fig. 4.12b. Likewise any measurement technique which averages over time, such as the integrating rainfall recorder, induces a 'moving-average' phenomenon in the data. When we smooth out the curve through lack of data or lack of precision in the data, such as fluctuations in the height of Pleistocene sea-levels, we automatically alter the correlogram and spectra derived from the data.

A fourth component of virtually all series is random fluctuations; these are the 'noise' elements which are independent of one another and form a 'background' to the trend, periodic and persistence effects already mentioned. In the autocorrelogram and the spectrum they are represented by insignificant fluctuations above zero. Effectively they represent the residual, uncorrelated and unexplained variation. Such random noise can also be explained by instrumental errors in observation or in timing of the observation which are irregular. From what has been said above, an instrumental error with a regular bias will produce persistence in the data rather than random errors.

To conclude this section, which deals with series for a single variable, it must be shown that by decomposing the series, i.e. breaking it down into the

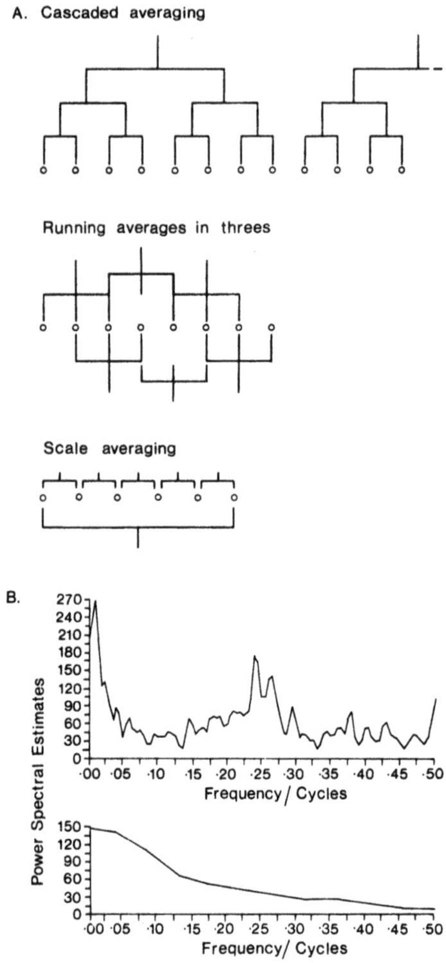

4.12 (a) Different averaging schemes for sampled data. (b) Scale effects of different averaging procedures, illustrating the smoothing effect of progressively longer scale averaging, which is to shift the variance (vertical axis) toward the lower frequency values and to reduce its overall magnitude (Thornes 1973).

various components, our understanding of the basic controls may be considerably enhanced. A discharge series may exhibit a long term trend of, say, decreasing run-off. This could best be observed in the mean annual discharge figures; by averaging over 12 months we remove all the higher-frequency (lower-period) observations. This is technically known as filtering. For a 40-year record then we have 40 observations. A visual plot at this stage would reveal the downward trend in mean annual discharge with time. By plotting the regression line we have a measure of the average rate of fall.

The second step is to remove this long term trend from the data, taking it from each of the 14,600 (daily) values of discharge and producing the monthly means. A plot of the monthly means of the new series (with trend removed) might show a marked seasonal oscillation, for example a Mediterranean river will show high winter and low summer discharges. This seasonal component might be removed by obtaining a best-fit Fourier series. Again these effects are subtracted from the raw data. Another technique is to take the difference between adjacent values or even to take the difference of these differences. A third is to use 12-month weighted moving averages. At each stage we may examine the raw data plots, the autocorrelogram and the power spectrum for evidence of what is left. In the discharge data we might have persistence components of, say 2–3 weeks. These are removed by fitting a low order autoregressive model, i.e. one that 'removes' the effects of short-term persistence. At each step as we remove successive components we are essentially identifying the sources of temporal variation in the data. Ultimately the objective is to have a set of residual points in which no serial component is left. This is then considered to be random noise.

It needs to be stressed at this stage that there are complexities and assumptions which have not been specified — these are covered by standard texts on spectral analysis such as Jenkins and Watts (1968) and Rayner (1971). The latter is a particularly lucid summary written by a geographer. Secondly, it should be noted that all the steps of the analysis will not always be necessary. Some data will be trend free, some will have no regularly fluctuating components.

Data from two or more series

So far we have compared different parts of the same series. Where two time series cover the same period and are measured over the same basic interval, then we can compare the two series. Consider fig. 4.13; for each series we may obtain the univariate acf and autocovariance functions. Additionally we may compare the value of series x at $(t + k)$ and series y at (t), so that the two series are lagged by k observations; in this case series y is assumed to lead series x. If we observed both series at value t, we would then have the equivalent of the ordinary correlation coefficient. If we have one series lagged behind the other we may obtain the equivalent of the single series acf and acvf (autocovariance function) which are, respectively, the cross-correlation and cross-covariance functions. For example, we might wish to compare frost heave in the ground with ambient air temperature using observations say every 15 minutes. It is reasonable to suppose that frost heave lags behind air temperature and to assume that more heave occurs as the temperature gets lower (though this is not necessarily the case). The cross-correlation might then show that after a lag of about 5 (75 minutes) frost heave responds to decrease in temperature by increasing (cross correlation coefficient is negative).

The cross-correlation function is a time (period)-based function; its frequency equivalent is the coherence function. This may be regarded as the correlation coefficient between two series at the same frequencies.

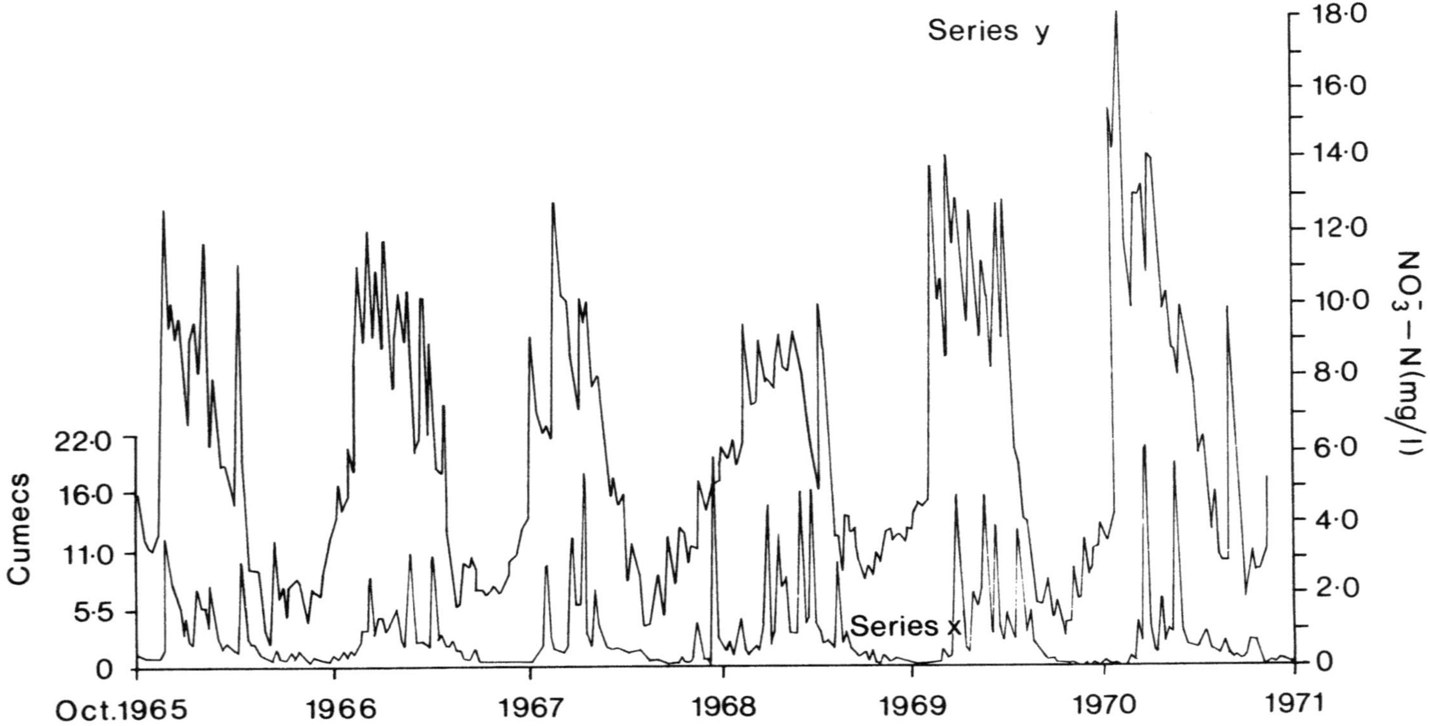

4.13 Weekly fluctuations in discharge and nitrate for the river Stour at Langham.

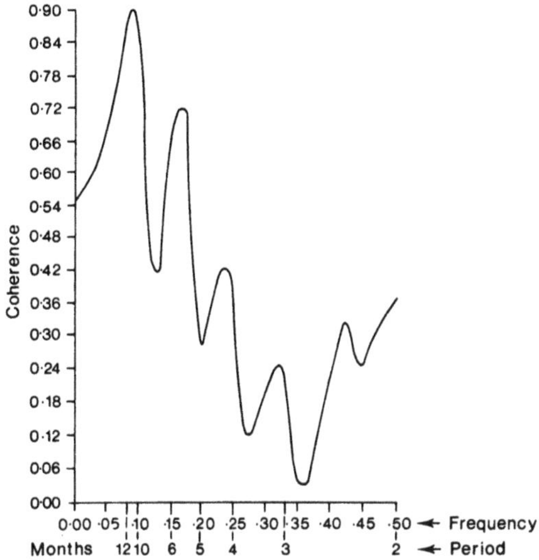

4.14 Coherence between dissolved solid load and discharge for the Animas river in Texas.

Figure 4.14 shows the coherence function for dissolved solid load and discharge for the Animas river in Texas. It will be seen that the annual frequency of the two series has a strong coherence. Where the two series are in phase the coherence will be high, where they are out of phase it will be low. A check on the amount of phase difference for various frequencies is provided by the lag or phase diagram. The phase value (normally measured in fractions of a circle and specific to each estimated frequency) will be zero when the two series are completely in phase.

As with univariate data, because we are dealing with sampled data, in *all* the plots of functions there exists the possibility of chance fluctuations in the plotted values. To identify the significance or otherwise of any point on any of the plots it is necessary first to construct the significance levels for the variable under consideration. Thus it is possible to say whether a particular value of the acf is or is not significantly different from zero. This is obviously quite important for determining the length of the persistence in a series. The acf may drop below the level at which it is significantly different from zero some time before it actually reaches the zero line. As with all statistical techniques careful consideration of the assumptions and limitations is necessary before application.

The transfer function approach

In the examples of the previous section, the comparison of two parallel series were considered. In the final section of this chapter, series which are themselves arranged in series are considered. They may be regarded as inputs

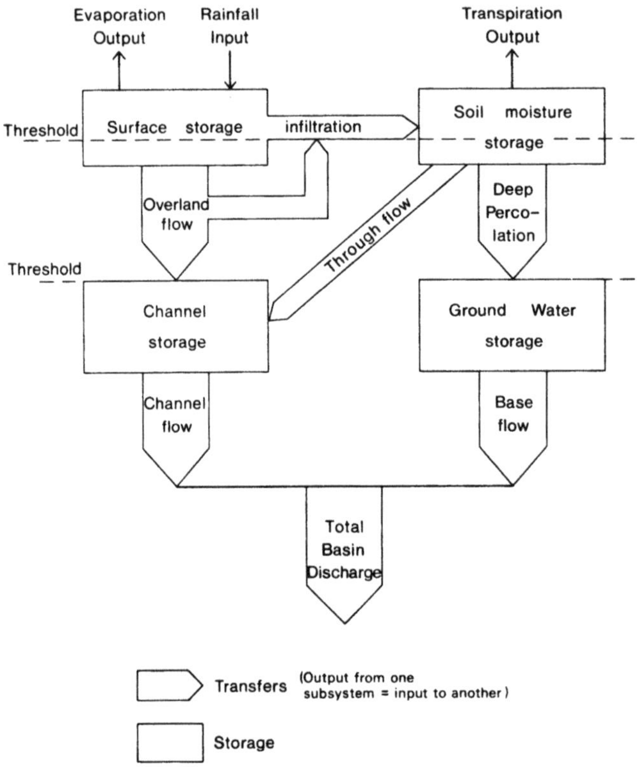

4.15 Hypothetical diagram of cascaded input-output systems involving thresholds.

and outputs to systems. Some examples are shown in fig. 4.15. Conventionally, in geomorphology these have been considered as process response models. Another way of viewing the systems however is to consider the mathematical function which relates inputs and outputs. Again hydrology provides us with a useful example. If we regard the drainage basin as an operating system we may expect to find a relationship between input (precipitation) and output (discharge). If we have this relationship we could forecast the discharge given the precipitation. Such a relationship is known as a transfer function.

The hydrologist Sherman (1932) proposed the unit hydrograph as the appropriate basis for creating such a transfer function. His idea was that, given a particular amount of rain in a particular basin, it could be disposed of across time by a standardized or unit hydrograph (fig. 4.16a). One inch of rain is spread across time by the hydrograph. If 2 inches fell the ordinates are doubled. If on the other hand a 2nd inch of rainfall succeeded the first, 1 unit of time later, the 2-unit hydrographs, staggered by one time unit, would be added together as shown in figures 4.16b-c. The unit hydrograph is

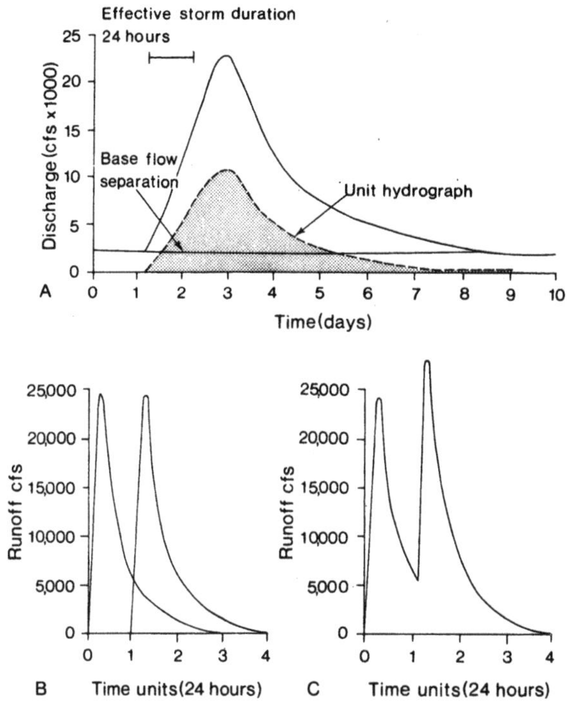

4.16 Application of the unit hydrograph method. (a) To a single storm. The shaded area shows the runoff produced by 1 inch of rainfall over the whole catchment. The outer curve is the hydrograph for a rainfall of 1.9 inches produced by multiplying the ordinates of the unit hydrograph. (b) and (c) Two successive storms showing summation effect.

equivalent to the impulse response function, for it gives the distribution of a single impulse (one inch of rain) in a unit of time across the system. We see now that the transfer function which comprises two parts for the output can be expressed as:

$$Y_t = gX_t$$

where g is the impulse response function, Y_t the output series, X_t the input series. The latter is sometimes called the forcing function, since it 'forces' the impulse response function to yield the output to the system.

Considered in this way, the transfer function is seen to be a 'discrete' expression, i.e. it relates to fixed units of time. Once again it can be considered in the frequency mode. It is known as the frequency response function. It is essentially a plot of the output as it would be affected only by the input function. It tells in effect what the system under consideration 'does' to the various frequencies. The idea may be illustrated with reference to a man-made reservoir built for flood protection. Floods are low frequency events, i.e. they occur relatively rarely. In fact the flood is often described in terms of its probable frequency; the 1,000-year flood for example. An effective reservoir is one which suppresses these low frequency events (low flows as well as high flows incidentally) so that only flows in the intermediate frequency, the average events, pass through the reservoir. Thus the frequency response function will be high for the intermediate and high frequencies and low for the low frequencies. In this way the reservoir is designed as a 'filter' allowing only the medium and low flows to pass through.

Although transfer function modelling has not yet been applied widely in geomorphology, it seems destined to become so, for geomorphologists as well as hydrologists are concerned with storage. This is especially so since deterministic models in differential equations have made use of the transfer function approach. For example, the Bagnold sedimentation model could be regarded as a transfer function model, determining as it does the likelihood of transport of sediment through a reach. Here the forcing function would be the input series, the output the sediment yield and the response function the transport-sedimentation relation for an impulsed input. A general assumption of the early work using hydrographs was that the transfer functions are linear, i.e. that given a series of fixed impulses their effects were additive, each unit impulse being equally important. There are good reasons why this should not be so, the first inch could be used, for example, for replenishment of soil water storage. This is being overcome by having non-linear response functions and/or by using several response functions by decomposing the system.

In this chapter an attempt has been made to introduce the techniques of time-series analysis which are widely used in other branches of science and industry. Awareness of these techniques will not, alas, produce the data which they require. On the other hand it will, hopefully, make the reader more aware of the care needed in designing the experiments in which time-based data is collected. It should, furthermore, increase the reader's awareness of the very inadequate nature of the published data on the rates of operation of geomorphological processes to be discussed in the following chapters.

5 Rates of operation of geomorphological processes

Geomorphologists are interested in rates of operation of processes for three main reasons: in the testing of models, in the provision of parameters and inputs for models and in practical problems of applied geomorphology. Process rates lie at the crux of some important models. A classical example was Davis' model of rapid uplift succeeded by a prolonged period of still stand during which the land was worn down by denudation. In a conflicting theory, W. Penck argued that denudation might keep pace with the rate of uplift, or the varying relative rates of the two would produce different suites of landforms. In an attempt to resolve this discrepancy Schumm (1963a) and Carson and Kirkby (1972) assembled data on rates of uplift and denudation and compared the two, concluding that uplift occurred very rapidly compared with denudation, at least as far as the available sample of rates showed (table 5.1).

Another commonly-held central model is that climate, through varying rates of operation of processes, controls the landforms of any region. This belief is extended to argue that the succession of landforms in a region reflects fluctuations in the relative dominance of different processes through time, as a result of climatic changes. Attempts to clarify the role of climate in relation to rates include discussion of the relative significance of the various processes in one climatic region; the way in which any one process varies under different climatic regimes and the net effect of all the processes as reflected by some 'integrating' measure in the different climatic regions.

The first approach is exemplified by the work of Rapp (1961) in Arctic Sweden and Leopold, Emmett and Myrick (1966) in the semi-arid United States. Both papers showed strong dominance of one process in terms of the work done. The second approach is exemplified by Corbel's (1959, 1964) attempt to compare the rates of calcium carbonate solution and total erosion (table 5.2) under different climatic zones and other attempts to study the disintegration of a particular rock type under various climatic controls. Another example is Troll's (1944) study of latitudinal variations in periglacial processes from the Poles to the equator. Studies on the integrated effects of

Table 5.1a Rates of uplift estimated for orogenic and isostatic conditions (from Schumm 1963a and Carson and Kirkby 1972)

Type of uplift	*Area*	*Rate m per 1000 yrs*	*Source*
Orogenic	Japan	0.8—75.0	Tsuboi 1933
	California	4.8—12.6	Gilluly 1949
	Persian Gulf	3.0—9.9	Lees 1955
	Southern California	3.9—6.0	Stone 1961
Isostatic	Fennoscandia	10.0	Gutenberg 1941
	Southern Ontario	4.8	Gutenberg 1941

Table 5.1b Rates of denudation estimated for drainage basins (from Schumm 1963a)

Drainage basin size	*Rate, m per 1000 yrs*	*Source*
1000 km^2		
0.0003	12.6	Fed. Inter-Agency River Basin Comm. 1953.
0.0030	2.55	Flaxman and High 1955
0.0800	0.06—0.22	Langbein and Schumm 1958
3.9000	0.03—0.10	Langbein and Schumm 1958
37-3,280	0.03—0.06	Dole and Stabler 1909

climate in producing rates of sediment yield include the work of Fournier (1960, 1969), Holeman (1968) and Stoddart (1969). Fournier's work is interesting in showing the importance of relief as well as precipitation amount and intensity on sediment yield. Schumm (1965) tried to bring together the two aspects (present and past) of climatic geomorphology in describing a model which is based on empirical data for the present time but which allows, under certain assumptions, estimation of the effect of temperature and effective rainfall variations on sediment yield in the past.

In these examples attempts are made to solve global problems. Rates are also sought to help in evaluating controls on the operation of a particular process. In so doing, the assumption is usually made that correlation or autocorrelation between the measure of the process and the 'controlling' variables reflects the relative significance of the variables in the physical mechanism which relates them. Thompson (1964) used this technique to elucidate the controls on gully growth, obtaining the rate of gully head advance in terms of the drainage area, slope of approach channel, a precipitation variable and the clay content of the eroding soil. In this case none of the other variables are time rates, though this is often the case. In studying suspended sediment load through time, an explanation is found in terms of a lagged discharge effect.

Table 5.2 Rates of erosion under different climates for mountain and plain areas of the world. After Corbel (1964) (Rate in m³/km²/yr)

	Arid (<200mm)		*Normal (200–1500 mm)*		*Humid (>1500 mm)*	
Climate	*Mountains*	*Plains*	*Mountains*	*Plains*	*Mountains*	*Plains*
Hot	1.0	0.5	25.0	10.0	30.0	15.0
Tropical	1.0	0.5	30.0	15.0	40.0	20.0
Extra tropical	4.0	1.0	100.0	20.0	100.0	30.0
Temperate	50.0	10.0	100.0	30.0	150.0	40.0
Cold	50.0	15.0	100.0	30.0	180.0	—
Polar	50.0	15.0	100.0	30.0	150.0	—
Glaciated polar	50.0	—	— 1000.0	—	— 2000.0	—
Glaciated non-polar	—	—	—	—	— 2000.0	—

The relationship here is complicated by a hysteresis effect (see below p. 98). An even more complicated case is the rate of lowering of a hillside slope in terms of the change in channel gradient at the foot of the slope. Here the difficulties arise from the inbuilt lags and feedback mechanisms which control the rate of adjustment of one to the other; the hillslope must wait until 'news of the change in gradient is brought to it', and until weathering allows it to respond. In addition to evaluating model or submodel relationships, rates are observed for use as inputs to models whose internal operation is assumed to be correct, or is to be tested. The use of input rates of operation is a common practice in hydrology where historical records, say of precipitation rates, are input to test models of catchment behaviour. Hindcasting wind data to produce wave refraction diagrams is a common example for beach process models. Alternatively, sample observations of rates can be used to generate statistical distributions of the rates which are themselves input as process generators in the models (cf. chapter 8). Very little use of synthesized rates of this type has been made in geomorphology and it has been mainly restricted to supplying relative frequencies in Monte Carlo simulations. Specification of rates as parameters is also important in deterministic models and a good example is the use of a constant flow rate in a model for channel discharge by Calver, Kirkby and Weyman (1972).

Finally output rates from models may be used for evaluating the success or otherwise of the model involved. Ability to reproduce a known and observable sequence, subject to specified and acceptable errors, is usually taken as a mark of success. Rana, Simons and Mahmood (1973) develop a model for the rate of change of sediment size by sorting along a stream channel and the output from the model is compared with rates observed in natural channels. The logical extension of this procedure is to adjust parameters inside the model until the output rates are 'right', a procedure variously known as calibration or optimization. This procedure, whilst often being desirable in a predictive situation, is less than desirable in explanatory research, for in overspecified models (those having too many parameters) a *range* of combinations of the parameter values may yield comparable output rates of operation of the simulated processes.

The third area in which observed rates are used is their direct application in the field of engineering and management. One need not seek far for examples; reservoir siltation, beach aggradation, cliff recession and soil erosion are crucial problems and study of rates, as well as the controls on those rates, is essential. There is, however, a certain irony, for while the pragmatic demands on the data on rates are such as to involve high penalties for providing the wrong rates, there is usually a sense of urgency that precludes long-term investigation. Nowhere is this more evident than in channel morphology where need to estimate rates such as bed-load transport and meander shifting, while being the subject of sophisticated theory in the text-books, is the subject of rule-of-thumb and high 'safety' factors in the real world.

In the general case, data has been collected for purposes that are far removed from the objective to which they are applied. This is essentially because data are expensive to collect, because the objectives of the investigation have been poorly specified or because the right variables were measured but in the wrong manner or at incompatible frequencies. Setting up universal schemes of data collection, though necessary if we are to have a general source of information, suffers from the inherent weakness that by the time the data have been collected model-building has progressed far enough to define a new conceptualization of the process which may require different data. Until geomorphologists have better generally based models, it will be sensible to collect data to test specific relations within the currently available models. The 'all-embracing' data set does not exist. At the same time more effective use will have to be made of existing data on the rates of operation.

General propositions

The concept of process rates in geomorphology implies the change in an observed variable, which is designated to measure a process, by reference to another variable or a reference scale. The variable chosen might for example be the trajectory values of an object in space and time or, more loosely, the net result of all such movements, in terms of a large-scale change such as the amount of lowering of a landform in geological time. Thus we may measure the process itself, or the result of that process, though in practice it is often difficult to do both (table 5.3).

Studies of rates must state the temporal reference scale employed. This is expressed by the fundamental dimension time (per unit of time) where the unit may be a second, month, year or a thousand years. The length dimension may be directly employed, for example one metre per second, as in velocity or uniform change (L/T) and in the velocity-time curve or acceleration (L/T^2). Occasionally unit length is used in order to compare phenomena of differing scale. 'Unit-bed width' of channels or $cm^3/cm/year$ for soil creep are common examples.

Rates may be expressed in terms of one variable which is chosen for one of the following reasons: to satisfy the objectives of the study, to provide a

Table 5.3 Example of the types of variable used to measure erosion rates. In this case possible measures for the denudation of a limestone area are shown.

	Method	*Example*	*Units*
Measurement	Direct measurement of rate	Solution rate	ppm/day
↓	Indirect measurement of rate, i.e. measurement of the 'result' of solution e.g. lowering of limestone surface	Weight loss Diameter loss Height loss	g/day mm/day
Inference	Lowering of limestone hill using solution load of stream at base of hill	Solute load	ppm/later converted to volume

sensitive measure of change or for logistic purposes. The success of choice depends on how responsive (range of variable) the variable is and how truly it represents the process. Typical examples of such variables are weight loss, change in percentage quartz or grain size distribution to measure weathering. Note that the specific process (e.g. oxidation) is not mentioned, the general term weathering is used instead. In a similar way we use sediment yield to infer regional rates of erosion. In this case, however, we may be many process steps 'removed' from the original generation of sediment. Here the whole erosion-transport path involves perhaps weathering, mass movement, slope wash, bank erosion and stream transport. Each is capable of individual measurement, but are grouped together as 'removal' — the 'process' by which regional denudation is achieved. Indeed the lowering of a mountain and the sediment passing the mouth of a stream hundreds of miles away may be so far removed that we speak of 'proxy' variables from which we infer rates. In reality of course it means that our measurements are becoming increasingly crude and our understanding of the 'process box' increasingly poor.

Process rates themselves vary through time. Whether a process is assumed to be operating at a constant rate depends of course on the scale of observation. Over a period of a year, for example, the result of isostatic rebound on tidal gauges may be assumed constant; over several thousand years the uplift rate varies enormously. Where processes do vary in the time scale under consideration, all the sampling and analytical problems outlined earlier (ch 3, p. 61) apply. Insofar as the product of the frequency and magnitude of these processes is a measure of the work done, it is particularly important to establish the relative frequency and magnitude of these rates.

In common with other variables measured in time, the rates may be continuous or discontinuous at a particular scale level. This may take two forms. A rate may fall to zero because of changes in the controlling variables and the process actually ceases to operate; a case in point might be sediment yield in arid zone rivers. In other cases the process is operative, but at a barely perceptible rate until some threshold is passed, as in certain cases of mass failure.

In cascaded processes the rate of one process may be conditioned by that of an 'earlier' process, thus hillslope wash and creep may be limited by the rate of weathering. In other cases, the rate of operation is conditioned by other variables which are themselves not the results of geomorphological processes.

This is a usual cause of spatial variations in rates, for example weathering is constrained by rock composition, and glacial erosion rates by pre-glacial configuration (which is not itself changing to affect erosion rates). In these cases there is not a dynamic cascade and the constraints on rate of operation are passive. In yet other situations rates may co-vary as a result of some joint cause and may even be in competition, as in the case of erosion and progradation in a delta. Finally we must remember that rates which have a functional dependency may be multivalued at different times. Such is the case of suspended sediment discharge with respect to water discharge. Because of hysteresis effects for a given discharge, the suspended load may take on two values, one representing the rising limb, the other the falling limb of the hydrograph.

Measurement of processes in experimental situations

Rates of operation of processes are measured by either observing the outcome of some process such as the change in a variable or using known laws to obtain the rate. In the first case we observe the trajectory of a body through space and use the distance travelled as a measure of the applied force; in the second case we obtain a force as the product of mass and acceleration. In both cases the process is the application of a force. When a rate of operation is expressed in length and time its property is described as *kinematic*; when mass is involved the situation is described as *dynamic*.

The simplest level at which processes can be measured is in the ordinal scale (cf. Krumbein and Graybill 1965). This is used when rates are arranged in some order of rank. For example we might class mass movements as very slow (1), slow (2), medium (3), fast (4) and very fast (5). Successive numerals on the ordinal scale are not equally spaced, and the difference in rate between (1) and (2) is not necessarily the same as that between (2) and (3). Obviously the level of information here is extremely low. Apart from counting the number observed in each rate-class, the only other useful information relates to the class in which the medium value (50th percentile) falls. The median value for mudflow rates might fall in the 'fast' class, that of quick clay flows in the 'medium' class and that of soil creep in the 'very slow' class. An example is provided by Varnes (1958) (fig. 5.1).

Most process information is measured, however, on the ratio scale, and it is the objective of most experiments to obtain data on this scale. The data here are again always 'derived' since time is always involved together with another dimension.

A more important division is whether the data has been obtained by direct measurement processes or by inference. In the first case the problems are essentially those of random or biased error of measurement. In the second problems of omission and dating also arise. Direct measurement has been present in fluvial studies for many centuries, both in the natural environment and in artificially-created hardware models. The last few decades have also been maked by a substantial increase in studies of rates in other areas, such as slope studies, glacial processes and weathering. Process rates studied in the artificial environment of the laboratory present a rather special set of problems related to scaling. The procedures have intuitive appeal because of the apparent ease with which the variables regulating the rates may be controlled. However, as King (1972) points out, unless the experimental results can be applied to the real-world counterparts they are of relatively little value. This is especially true where a change of scale alters not only the rates but also the mechanisms.

Attempts to overcome these scale problems involve geometrical, kinematic and dynamic similarity between the model and nature and the use of numbers which are dimension free, such as the Froude number in hydraulics. In geometrical similarity the ratio of lengths (λL) in the prototype (Lp) and model (Lm) are the same, and this is expressed by $\lambda L = Lm/Lp$; likewise we may have ratios in time (λT) and in mass (λM). Kinematic similarity implies

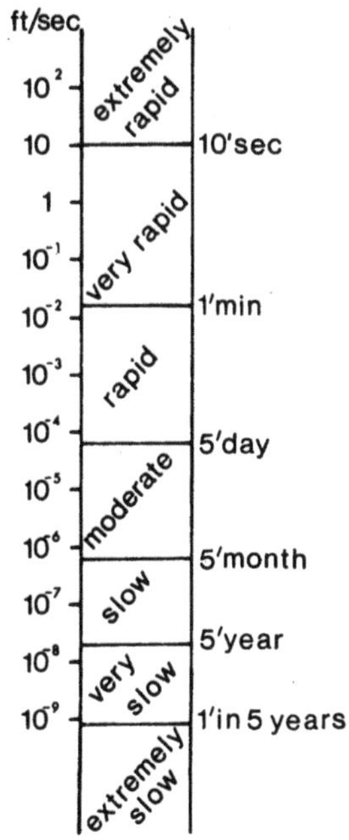

5.1 Example of ordinal and interval scales for classifying mass movement (after Varnes 1958).

that models have fixed $\lambda L \lambda T$ and in dynamic similarity $\lambda L \lambda T$ and λM are all fixed proportions between the prototype and the model. Two things of particular importance have to be noted here. First, true dynamic similarity is impossible to achieve unless the prototype and the model are the same size. This arises from the fact that λg, the proportionality factor for gravity, is always unity. Attempts to overcome this (Schofield 1971) have involved the use of large scale centrifuges, where scale models can be tested for various g values. The aim is to scale gravity to the model size. Models of slope embankments and landslides have been effectively tested in this equipment. In hydraulic models, the use of water frequently implies that the viscosity and density proportionalities will also be equal to one. This usually raises the second important point: that it is not essential for one scale to determine all the relationships. In distorted models used in hydraulic engineering the depth proportionality is frequently different from that of widths; once two scales are fixed, however, the rest usually follow. The ratio between the fixed scales is the distortion ratio of the model. Dynamic similarity implies proportionality between derived units (such as bed shear stress) as well as fundamental units in the model and the prototype. Finally, and of particular interest in this context, dynamic similarity implies a fixed time ratio which in turn suggests the possibility of forecasting from the model.

Geomorphological process rates have not infrequently been studied in scale models. Because of the slow rate of most processes effort has been by and large directed towards speeding up the model by making $\lambda T >> 1$. A good example is the rate of mechanical breakdown of various rock types under the action of frost, notably the work of Wiman (1963) following earlier work of Tricart (1956). The experimental procedure of both workers involved temperature changes in freezing rooms and the time scales involved an Icelandic type (high frequency) oscillation and a Siberian type (low frequency) oscillation with 1 cycle per day and 1 cycle per 4 days respectively; the processes allow therefore for speeding up by a factor of x90 in the second case. A danger here may arise if the 'restoration' period is inadequate, i.e. if a period in which a new equilibrium is reached is itself truncated by entry into the next cycle.

Model situations are also used to study rates in situations where it is impractical or uneconomic to wait for nature, even though the time scale may not be 'collapsed'. This is the case in laboratory and field attempts to study the effects of precipitation on the rate of soil erosion. The dimensional problems outlined above are common here and the scaling problem becomes crucial in the desire to obtain appropriate impact velocities for falling raindrops since their momentum is considered highly relevant to surface erosion rates. Artificial rainfall simulators are also used in the open air for the same purposes though the most useful of these have obtained either exponent values for existing mathematical models or better understanding of the mechanics of the processes involved. Emmett's (1970) paper has both of these objectives and is illustrative of the experimental approach to the rates of operation of geomorphological processes.

It is perhaps above all in the laboratory flume that some of the most sophisticated experiments on rates have taken place; this is possible only through the high degree of control and careful attention to scaling problems which are common in the best of these experiments. Of particular interest have been studies of the rate of sediment transport, and especially that of bedload since, while the theory of bedload transport is well endowed with formulations as to the mechanism, the accurate measurement of bedload in stream channels has proved particularly elusive. Even in the flume there are considerable difficulties in separation and measurement of bedload, though they tend to be less than in real channels. In such studies, the usual procedure is to establish equilibrium conditions by adjusting the tailgate of the flume, obtaining uniform flow and then allowing the bed slope to become stabilized. Sometimes the time required to reach such an equilibrium may take many days. The sediment transported is usually trapped in a collection box and then weighed. Sediment transport thus measured is then compared with stream power, shear stress based on hydraulic radius or depth and various other measures of the strength of flowing water. An example of such a study is that of Williams (1967) who showed the considerable importance of maintaining constant water depth if a unique relationship between sediment transport rate and the above factors is to be obtained.

Direct measurement in the field

It is not our objective in this section to cover the many and diverse field measurements of rates, but rather to select a few which reflect the trend in recent years towards this type of observation. One of the most coherent developments in this respect has been the instrumentation of catchments along internationally specified and agreed lines under the general guidance of the World Meteorological Organization. Three kinds of hydrological catchments seem to have developed, those which are unchanging in their principal hydrological factors (cover type, drainage density etc), those in which observations are made throughout normal changes (such as urbanization) and those in which specific changes have been induced for experimental purposes. By and large studies have concentrated on hydrological rather than geomorphological processes though there are notable exceptions to this (Gregory and Walling 1973). A major problem for such studies has been the rapid change in conceptualization of basin processes with the attendant decisions as to whether to scrap an existing expensive installation in favour of some new experiment or whether to continue. A second problem relates to the decision as to whether one should adopt a philosophy of a large number of modestly instrumented catchments or a few very well instrumented ones, since the two are often in financial competition. Gregory and Walling (1973) point out that factors of cost, leakage, representativeness, the problems of comparison and the difficulties of detecting change have been raised as criticisms of experimental watersheds, though in fact their book and the study of small catchment data by Gregory and Walling (1974) are adequate testimony to the utility of such studies.

In geomorphological terms the pattern and timing of discharge is obviously an important controlling variable and in particular the relative contributions made by the various components to the hydrograph by overland, through-flow and ground-water flow. This is because discharge is a key variable in many other process rates; sediment transport, solution rates, soil erosion rates, bank collapse and so on. Many of these variables have been measured successfully and with relatively little difficulty in controlled situations. Some, however, although of crucial importance, are notoriously difficult to measure, and this includes sediment transport, changes in bed slope, changes in channel roughness and flow characteristics of mountain streams.

In terms of sediment load sampling the need is threefold: (i) general information on a network basis for applied needs, (ii) special information at specific problem areas for water management (reservoirs and power station intakes, for example), and (iii) data for understanding the relationships between water, sediment and the environment. Specialized catchment studies have generally been geared toward the latter. Bed material sampling presents the greatest difficulty and yet is perhaps one of the most crucial forms from a geomorphological point of view. The techniques range from repeated hand sampling and measurement at a station (see Kellerhals and Bray 1971) to techniques involving measurement of sonic intensity under water (Hubbell 1964). Bedload is difficult to measure for several reasons. First, any mechanical device placed near the bed will disturb the flow and hence the rate of movement. Secondly, the sediment movement and velocity of water close to the bed vary considerably in respect to both time and position so that theoretically at least the mean at each point for some fixed time period, rather than a 'spot' observation, ought to be recorded. Finally, because of very considerable variations in the cross-sectional distribution of bed material size it is necessary to be able effectively to trap all particles moving along the bed when and if they pass over an area of the bed in a specified period of time. The actual form taken by bedload samplers has been described by Gregory and Walling (1973) and the main types are pit traps, slot and tray samplers and basket samplers. Hubbell (1964) gives levels of efficiency for these various techniques which are worrying to say the least. For box and basket he estimates average efficiency to be of the order of 45%, while slot techniques were shown to have efficiencies of between 38% and 75% depending on the velocity.

In relation to sediment transport, there is the question of rates of scour and fill. A more diverse set of techniques is used here although the basic problem of sampling in a water-bearing stream is usually present. One simple procedure is to make repeated cross-sections and then obtain successive areal differences. This has to be integrated over the length of the channel reach to change area into volume and it is found that channel elevation changes quite rapidly in space, so that the degree of approximation depends on the spacing of the sections as well as the frequency of observations. As Colby (1964) points out many of the observations are made at sites chosen originally for their suitability for stream gauging. Oviously data collected on water surface width basis is inadequate since width fluctuates very rapidly. Insertion of plastic ribbons or chains into the bed of ephemeral streams has been used by several workers to determine the depth of maximum scour, but the problems here is that the time of maximum scour may differ along the channel and even in the same section.

Finally, the measurement of regional rates of erosion (table 5.1b) should be briefly mentioned, because they present the only data for world-wide comparison of rates under different climatic and lithological environments. These are derived from sediment and solute loads to give the volume of material removed per unit area within a given time on the basis of some assumed density of the material. Variations in the specific density (especially considering that soil is being eroded), the effects of human activity in the basin and the inconclusiveness of such rates when extrapolated all lead to serious questions about the utility of such information. Actual rates of sediment yield seem preferable though of course these too have all the attendant problems of time and space variations.

In *all* rates of operation in fluvial processes, we must seriously consider the utility of contemporary measurements in the role of longer term evolution of the landscape. The dictum that the present is the key to the past is perhaps more doubtful and yet more critical than in any other area of geomorphology. Insofar as a knowledge of rates is essential to the study of mechanisms in the contemporary environment, there can be little criticism. The main objection is the extrapolation of arguments about the relative importance of different rates. This is because (i) the past environments are known to have been highly variable in time and space, (ii) present estimates of many rates are by and large poor in quality and have high standard errors, (iii) we have little evidence to support the assumption that such rates are the direct response to immediate processes, given the possible transitory nature of the system state because of (iv) the widespread, almost ubiquitous impact of human activities.

In the study of rates in slope morphology and evolution these problems are no less critical when trying to establish relationships between form and process. However in this instance, greater attention has been paid at least in recent studies to process-mechanisms rather than to process-form relationships. The books by Young (1972) and Carson and Kirkby (1972) both give good reviews of the technical problems of assessing rates and we shall refer here to only two examples, the determination of flow rates within the soil, and the assessment of rates of soil loss from the surface.

Flow rates beneath the surface are important both in terms of understanding the hydrograph, in relation to the nature and development of subsurface structural features, and in terms of subsurface solution. This work was initiated by Whipkey (1965) and has since been carried out by many others including Dunne and Black (1970) and Weyman (1970). The basic proposition is that various horizons in the soil are intercepted and the water collected by the intercepting mechanism is drained off and measured. Knapp (1973) has outlined in detail the procedures involved and draws attention to the two basic problems which are (i) effective monitoring of the throughflow,

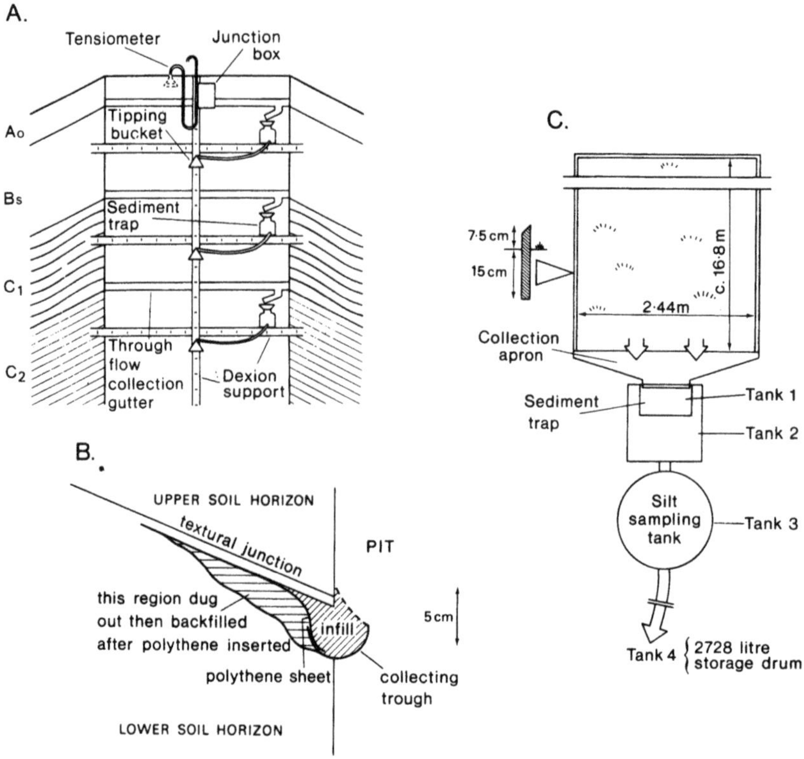

5.2 Examples of instruments which might be used to measure throughflow and soil wash in the field. (a) and (b) Details of a field installation to catch water and sediment from the throughflow of successive soil horizons (after Knapp 1973). (c) Soil erosion and surface water runoff (after Gregory and Walling 1973).

and (ii) minimization of the effects of measurement on the natural flow. Normally the technique consists of a trench or pit dug into the soil with the collection of water from the upslope face at selected depths by open rainwater channelling (fig. 5.2a). Sometimes the flow is provided artificially by simulated rainfall. To help flow into the receiving monitor the lower surface of the soil horizon is led into the trough by plastic sheeting (fig. 5.2b). The water led away from the pit is measured by some device such as a tipping bucket gauge.

The measurement of rates of surface wash is normally in terms of trough collection (fig. 5.2c), elevation changes on marked rods (fig. 5.3) or tracers. The last two are measures of point loss and particle movement per unit of time respectively; the first is volumetric. The choice and location of profiles depends on the study in question but the International Geographical Union has attempted to outline the main considerations in this respect (Leopold and Dunne 1971). Troughs for collecting sediment on the slope normally consist

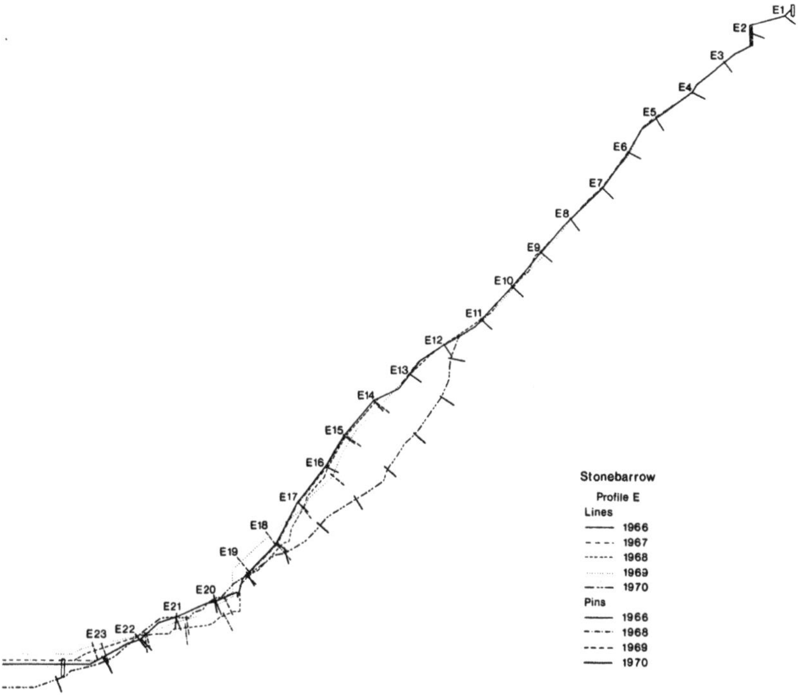

5.3 The use of erosion pins to measure the rate of degradation of a rapidly evolving clay slope (after Brunsden 1974).

of an elongated metal box into which the slope sediments may move across a normal surface, and from which soil loss may occur without loss of sediment. The very existence of the trough tends to accelerate the loss though this may be offset by an apron of concrete. Use of erosion pins is now quite widespread and the results for a survey of the backwall of a mud-slide are shown in fig. 5.3. Steel rods a metre in length and a half centimetre in diameter are inset and the degree of exposure measured on the stakes. Sometimes a washer is placed on a stake so that the maximum amount of loss preceding a depositional phase can be recorded. This technique was used by Leopold, Emmett and Myrick (1966). Finally, attempts have been made to evaluate the rates of process operation on slopes by observing the amount and direction of movement of marked stones or radioactive tracers. Schumm (1964), for example, observed the amount of movement in each size class and multiplied this by the amount of volume of each size-class present in the slope deposit. The rate of downslope movement of stones of various size by rainsplash according to size and slope angle studied is similar in approach.

These few examples yield but a modest indication of the large range of direct attempts to measure process rates. They illustrate the principal difficulties involved, of sampling, instrumental and human errors, and they involve mainly studies of process mechanics rather than attempts to obtain long-term change. They are all examples of situations in which the processes take place fairly quickly and in amounts which are measurable. The more difficult studies of contemporary rates are those involving very slow rates, such as soil creep, in which the magnitude of experimental error is almost as large as the rate under observation. Moreover in both slow and swiftly operating processes it is not always clear which process is responsible. For all this, actual observed rates are for the most part superior to those inferred, especially where process mechanisms are involved.

Inferring rates of operation

Where there is not sufficient time or equipment to measure rates directly, a common practice is to infer rates from the results of a process or group of processes. Commonly the actual processes or the locations of the process activity are imperfectly known and only general or 'average' statements may be made. The most usual application of the method is in estimates of regional denudation rates where we require (a) the source area of the resulting sediments, (b) the volume of deposits derived from this source and (c) the duration of sedimentation (which is assumed to be equal to duration of erosion).

Reservoir sedimentation has been used for this procedure (Subcommittee on Sedimentation 1953). Reservoirs are favoured because they are considered to capture a large proportion of the load, depending on their efficiency as a sediment trap. Certainly there are losses through spillage and these are mainly of dissolved load. There are also problems of coring through sediments to estimate the densities and consolidation rates of the various layers though these are overcome in part when reservoirs are occasionally drained. Several researchers have tried to express the amount caught in reservoirs as a ratio of the total sediment yield of the basin above the reservoir. This is called the delivery ratio and when expressed as a percentage is normally below a hundred. Delivery ratios depend on accurate estimation of the sediment yield above the basin, which is normally carried out by means of some soil loss equation. Figures of 5-60% are quoted, the amounts reflecting channel and slope storage potential which might be related to basin shape, slope and run-off. The estimation of delivery ratios seems to be of dubious value given the large potential errors in virtually all the measurements, but the idea of slope and channel storage affecting observed sediment yields is both correct and important. It is in large measure responsible for the difficulties of obtaining a formal expression of the relationship between climatic variables and sediment yield.

Rates have also been estimated from deposits as diverse as those of the continental shelf, the deltas of the Mississippi and the basin deposits of the Ganges and the Brahmaputra (Menard 1961). The order of magnitude of error in these studies is much larger still and arises from problems of age estimation as well as difficulties of obtaining the volume of sediments. In between fall the kind of estimates made by Miller and Wendorf (1958) based on volumetric measurements of terrace fill from the Teseque valley in the south-west United States. Here horizons were dated by C^{14} and archaeological evidence, the main difficulties being those of unconformable sequences. There is little or no evidence of cases of non-deposition. Similar efforts have been made to obtain information on the rates of accretion of flood plains in historic times.

Attempts have also been made to infer overall denudation rates from erosional evidence. The general proposition is that the volume removed is calculated for a known period of time. The difficulties of obtaining dates are more acute here and the problems are compounded by the need to reconstruct the original surface. The success of this reconstruction depends on the simplicity or otherwise of the landform involved. Ruxton and McDougall (1967) dated the Hydrographer's Volcano in north-east Papua by potassium-argon methods, obtaining a date of 0.56 ± 0.06 million years. The volcano is roughly circular in plan with a radial pattern of deeply incised consequent streams and its present form and reconstruction form before erosion were obtained from aerial photographs and generalized contours respectively. They assumed that the processes of construction were complete before erosion began and that there was an original symmetrical form of the cone. Subtraction of the two maps from sections (fig. 5.4) yields changes in erosion rate with height as well as overall rates. The practice of estimating the amount of erosion between uplift episodes is also common in denudation chronology, though here reconstruction is based on isopachytes and the assumption of construction completed before erosion commenced is more difficult to sustain.

Rates of specific processes are also inferred for the longer term. Dury (1964a and b) attempted to infer the past rates of run-off (and precipitation) from the hydraulic geometry of river channels. He showed that present-day meandering rivers sometimes flow on alluvial fills which related to valley meanders, and found that the wavelength of valley meanders was about 9-10 times that of stream meanders in lowland Britain. Using the relationship between meander wavelength and bankfull discharge he suggested bankfull discharges of the order of 50-60 times greater than these now existing. Another example is Ruhe's (1954) estimate of the rate of development of drainage patterns by analysis of the patterns developed on tills of different ages.

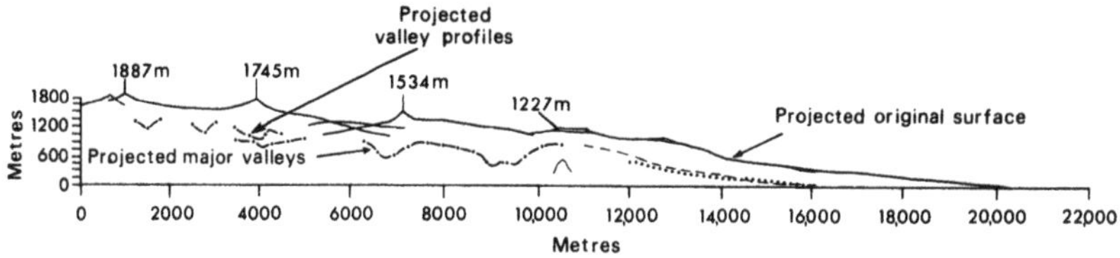

5.4 An example of an attempt to measure overall denudation rates from erosional evidence where volume removed in a known period of time is calculated from successive time levels of an erosion surface. Hydrographer's Volcano, Papua (after Ruxton and McDougall 1967). See text for full explanation.

In glacial geomorphology inference of operation rates of processes was especially important before intensive studies of existing glaciers became available, though here too certain phenomena, such as glacier retreat, have been the object of direct observation. Estimating rates of advance and retreat involves establishing distances between events known to be successional, such as the formation of recessional moraines, and from these rates inferences concerning climatic causation have been made. Retreat rates for relatively recent times have improved as a result of the development of better dating techniques (see chapter 2).

Rates of denudation

Rates of denudation are used in three areas of interest: (i) regional rates, designed to test macroscopic theories about evolution, climatic control on processes, and controls of various variables on sediment yield itself; (ii) comparative studies of the relative roles of different processes within particular environments to obtain a measure of the principal controls in process-response studies; and (iii) studies of the variations of individual processes through time with a view to obtaining the specific controls on these processes.

The first point in the use of regional rates of denudation is that the controlling variables are scale dependent. Thus on the world scale we expect relief and precipitation controls to account for much of the variance, while on the scale of several hundred square miles local ground slope, rock type and local vegetation cover would probably be more important. Even at this lowest level, however, especially in areas with strong seasonal climatic variations, precipitation rates remain important.

Global controls of relief and climate were demonstrated by Fournier (1960) who showed that a basic difference existed between high and low relief areas in the 96 basins studied, and that within this subgrouping precipitation seasonality was a most important control. A plot of the ratio of the square of the rainfall of the wettest month to the mean annual precipitation $(p^2)/P$ against sediment yield for different relief groupings was obtained by regression analysis and is shown in fig. 5.5. The two variables were also combined into a single empirical equation as follows:

$$\log F = 2.65 \log \frac{p^2}{P} + 0.46 \log \bar{H} \tan \theta - 1.56$$

F = suspended sediment yield/ann. (tons/kms^2/yr)

$\bar{H}$ = mean relief

θ = mean slope in drainage basin

This model has not been criticized in detail, and although some authors (Fleming 1969, Strakhov 1967) have modified the results and disagree on absolute values, the expected global patterns are much the same.

On the continental scale Langbein and Schumm (1958) have shown the impact of effective precipitation on sediment yield through its intermediary

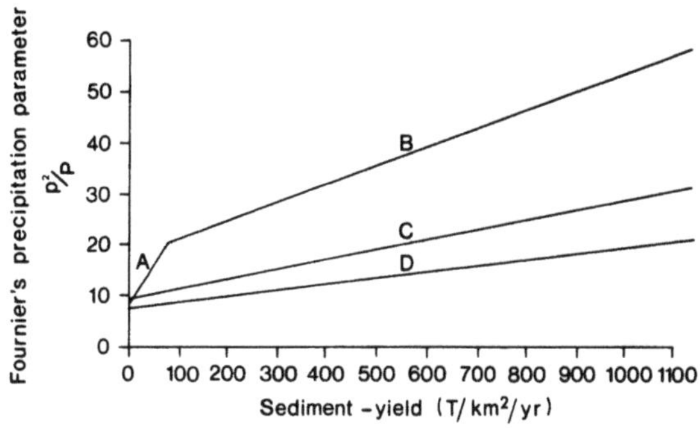

5.5 The relationship between precipitation amount and intensity and sediment yield derived by Fournier to distinguish different climatic regions (after Fournier 1960). A = Low relief, temperate climate. B = Low relief, tropical, sub-tropical and semi-arid climate. C = High relief, humid climate. D = High relief, semi-arid climate.

of vegetation cover. The continental data also shows an astonishing contrast between sediment yields for large and small basins, a difference which is related to differences in delivery ratios, to the dramatic impact of human activities on sediment yield and to the local effect of lithology.

On the micro-scale, also using suspended sediment variations *within* a single basin, the study of Lustig and Busch (1967) is of great interest. Variations in the basin reflected precipitation rates but also variations, both spatial and seasonal, in the infiltration capacity. They also draw attention to the importance of spatial variations in frequency, for example finding an increase in the frequency of low water flows in the upper and middle parts of the basin and a decrease in the lower part of the basin during the period 1960-1963. Geological and topographical controls mean that the one area in the basin produces material at a rate twice that of the basin as a whole.

The second general point about regional rates is that according to most general calculations, contemporary global rates are of the order of 7-8 times those occurring in the geological past (Menard 1961). This reflects the impact of human activities on soil erosion following deforestation and extensive ploughing (Brown 1970). Furthermore, during the Quaternary there were isostatic and eustatic changes of great importance which, coupled with present-day rates by systems lags, have led to increased erosional activity. Fairbridge (1961) makes the further points that changes in the extent of semi-arid and the humid-tropical zones in the past, and increase in the relative area of the continents through time, should be considered in any attempt to relate present rates to be long-term landform evolution.

The third general point is that empirically we can demonstrate the overall trends in the rates of denudation resulting from controls of climate, relief and lithology in a forecasting sense. However, if our objective is to understand and explain these controls we are forced to consider the individual processes in more detail because different kinds of processes operate at different rates in different areas.

Table 5.4 The relative importance of processes in a periglacial mountain climate (after Rapp 1961)

Process	*Volume* m_3	*T*/km²	*Average movement* m	*Tons moved per vertical* m
rockfalls	50	8.7	90–225	19,565
avalanches	88	15.5	100–200	21,850
earthslides	580	69.4	0.5–600	96,375
talus creep	300,000	–	0.01	2,700
solifluction	550,000	–	0.02	5,300
running water dissolved load	150	26	700	136,500

A classic study was undertaken by Rapp (1961) in which the relative significance of different processes in one area was investigated over a prolonged period of time. By measurement and spatial integration Rapp was able to present data on the rates of movement of various processes and the amounts of material which each had removed in a given period of time in the small mountain valley of Karkevagge in northern Sweden (table 5.4). The results show that, in order of importance, the major processes responsible for geomorphological change were (i) transportation of dissolved salts, (ii) earth slides and mass movement, (iii) dirt avalanches, (iv) rockfalls, (v) solifluction and (vi) talus creep. Jackli (1957) in a comparable study in the upper Rhine, in its alpine section, also showed the importance of dissolved solids.

Although Corbel (1959) had argued for high rates of solution of calcium carbonate in high latitudes the pre-eminence of dissolved load came as a surprise. In both cases the authors pointed out the limitations of their investigations and insisted that adjacent valleys might show quite substantial differences. This pioneering work has so far only been emulated by two other works, both valuable in their own right, if less detailed and for shorter periods. The first is the study of Leopold, Emmett and Myrick (1966) of the small Arroyo de les Frijoles, near Sante Fé in New Mexico. These results, like those of Rapp, are based on direct measurement, and involve the amounts of sediment produced by different erosion processes in various physiographic positions in the drainage basin. This paper is especially useful for the discussion of the error sources involved in what we earlier called spatial integration, that is the extension from point locations such as erosion pins to the whole basin. For example, in estimating the debris production due to soil creep, the original data consisted of erosion pins located in sloping gully walls. They argue that erosion processes should deliver material in proportion to the total channel length of a given area. This was estimated to be 70 miles (including both banks) and was multiplied by a figure of 28 ft³/mile of slope base/year giving a value of 1,960 ft³/year or 98 tons/year. The results of their investigation further revealed that in this semi-arid area 97.8% of sediment production could be attributable to surface erosion by sheet wash, the relative importance of gully erosion and mass-movement being negligible. One other study of this type by Slaymaker (1972) in the temperate climate of central Wales is important in laying rather more emphasis on the magnitude and frequency concept in relation to sediment amounts.

We turn finally to a few studies whose objective has been the study of a particular process and variations in its rate of operation over space and time. In some work this has been studied simply as the presence or absence of a particular process at different times. Such is the basis of the maps of Poser (1948) for example, who mapped the occurrence of fossil frost cracks and other fossil periglacial features, including involutions, throughout Europe. These reveal the former extent of particular processes throughout Europe in areas where they do not now occur. In a more quantitative study, Corbel (1959), in his study of the Karst of north-west Europe, endeavoured to show how processes of solution might vary regionally according to climatic control and locally according to variations in the principle types of limestones. Not all studies of a particular process have approached the problem from an inductive point of view; some, indeed, have approached it from a purely speculative point of view, usually in a crude attempt to relate processes to simple and, geomorphologically speaking, largely irrelevant climatic variables. More recent attempts to model deductively the role of various climatic controls on process are likely to meet with greater success. We are thinking here of Schumm's attempts to model river responses to climatic change (1968) or Carson and Kirkby's (1972) deductions of the control of precipitation on processes producing characteristic slope forms. Ultimately, any survey of the literature will reveal that the only area of geomorphology which has adequate data for the study of spatial and temporal variations of processes relate to fluvial processes and even here the scope is largely restricted to particular areas of the world, notably north-west Europe, North America, the Indian sub-continent and the Soviet Union. It is hardly surprising that, given the current inadequacy of data on contemporary rates, assumptions about past rates and the supposed form-responses can only be viewed with a reserve verging on suspicion.

6 Qualitative temporal models

The first part of this book has been concerned essentially with the collection and analysis of information about changes occurring in the temporal framework and the analysis of these changes. In the second, we turn attention to research efforts involving the creation of various types of models of the way in which events change in time. Models are essentially statements about reality made with various levels of abstraction from reality. These statements may be made mechanically, symbolically or verbally. A toy train may be regarded as a mechanical statement about a real train involving not only reduction in size but also a high level of abstraction. A regression equation, together with the associated correlation statistics, is a symbolic representation of the statistical relationships between two variables. The statement: 'The ice advanced approximately as far as a line drawn from Bristol to Finchley and then retreated again' is a very highly simplified model in the form of a verbal statement.

Models are very similar to theory and indeed some people regard all models as theory and vice versa. Harvey (1969) helps to clarify the problem by arguing that models fall short of being theory insofar as 'A model becomes a theory about the real world only when a segment of the real world has been mapped into it.' This mapping in of real world data almost invariably means in geomorphology the provision of numeric data in the form of parameters or variable values, point locations of objects and events in time and so on. Thus, as we might expect, models evolve into theory and qualitative models evolve into quantitative models which in turn may develop into theory.

Another source of confusion has related to the use of the term 'anologue'. Chorley (1964), an authority on the history and philosophy of geomorphology, regards all models as analogues. It seems preferable, however, to reserve this term for the subject of models in which the physical components of the model are completely dissimilar from the data or theory which they represent by their structural similarities.

It is perhaps more useful to consider models in terms of the type of function they perform because the multiplicity of definitions, of types, of formal procedures of model construction and manipulation are all closely related to the functions which they perform (table 6.1, after Harvey).

Table 6.1
A classification of models based on their function
D = data; M = model; T = theory; H = hypothesis; L = law
(modified after Harvey 1968)

Type	*Function*	*Example*
1	D → M → T,H,L	Correlation structure models. Horton's Laws.
2	T,H,L → M → D present	Allometric growth. Topologically random network theory.
3	(a) T,H,L → M → D future	Mathematical models of slopes.
	(b) D past → M → D future	Stochastic forecasting in hydrology.
4	(a) M → T	Deductive model-building. Random flight models.
	(b) T old → M → T new	Extension and/or restructuring of Darcy's Law.
5	T,H,L → M	The Davisian model, Valentin's model of coastal change.

(i) At the conceptually simplest level, models of relationships among collected data are used inductively in the discovery of a theory, hypothesis or law. Investigation commences with the analysis of a data set, often collected for some other purpose and the setting up of correlation structures which by their very character suggest particular hypotheses about the relationships between the variables. This type of work in the quantitative field was initiated by Strahler (1950) and Melton (1958), found its fullest extension in the study of hydraulic geometry of stream channels and is expertly reviewed by Chorley and Kennedy (1971) under the guise of systems theory. The transition from model to theory in this field is exemplified by the work of Schumm (1954). In the time-domain, the development of models of glacial advance and retreat, leading ultimately to theories of climatic change, has followed a very similar, if initially more qualitative path.

(ii) Another kind of use of models is in establishing the domain over which a theory or law might hold, working towards the data, rather than away from it. Put simply this means if the theory or law is applied over a set of real world values, what are the upper and lower limits of this application? An example of this we might follow is the Woldenburg model for allometric growth in a set of drainage basins (Woldenburg 1966), finding however that while the property holds for areas when there is spatial competition, it could break down for other properties such as stream length. Provided that the model adequately represents the theory, failure of the model to conform to the real world will usually indicate inadequacy in the domain of the theory, i.e. the theory does not apply!

(iii) A similar use of the model, which also works from theory or law to data via a model, is prediction. This kind of use is discussed particularly in chapters 7 and 8 and involves a quite explicit time component. If, for example, we adopt the physical laws of Newtonian mechanics and model from them what happens when a dam breaks, then, given the characteristics of the channel, the downstream time and space trajectory of the ensuing water wave may be predicted by solution of the characteristic equations. A very considerable number of mathematical models have been developed to predict the value of real variables under specified conditions, some of which are discussed in chapter 7. When these are set up as formal probabilistic processes, such as Markov or inventory processes, the models are called stochastic models. These and other types of stochastic models are considered in chapter 8. A special case is the so-called 'black box' or transfer function approach in which the model is represented by an equation, a statistical distribution or a frequency response function which translates 'input' into 'output', over time. This model may simply use past values as 'input', or it may take values developed by the model at any particular time and feed them back into the 'input' so that the 'output' incorporates the effects of this feedback. These differences are shown diagrammatically in fig. 6.1. A great deal of interest has focused on negative feedback models, and as yet little attention has been paid to positive feedback.

(iv) Two particular uses overwhelm all the others in importance. First, models are used in creating theory as intermediate stages in the process of deduction. Secondly, they may be used for the restructuring or extension of existing theory by additions or changes to the existing theory and its testing to establish new theory. The reason for the importance of this particular function of model building is that it either seeks to break new ground or builds creatively on what has gone before. There is very little 'new' theory in the real sense of the word but there are occasionally ingenious, lively, speculative and often spectacular leaps ahead in model-building which carry the subjects forward; sometimes into new theory, as with Bagnold's (1966) work on sediment transport. The extension of earlier theory by restructuring is well exemplified by Kirkby's (1971) formulation of Gilbert's (1877) continuity model for sediment transport. Here the restructuring not only lays open a formal system for testing the model against reality, but allows a more extensive development of it in terms of process.

(v) Finally, models may be used to convey the essence of a complex theory, often resulting in confusion in the recipient's mind between model and theory. Thus the Davisian model as commonly taught in schools represents but one of several possible cases of a poorly developed but nonetheless complex theory of landform development. Some would argue that in essence Davis' work never got beyond the stage of model-building (see p. 121). It remains indisputable however that almost every teacher of geography in the first half of the century saw in this, as in the models of Penck and King, frameworks for conveying the essence of the range of possible forms that macroscopic landform evolution could follow.

The types of function performed by models as described above overlook the fact that, of course, models sometimes perform several different functions well, several functions inadequately and so on. Confusion sometimes arises when a model built to satisfy one of these functions is also required to serve in another role. Models used inductively to progress towards a theory are often used as explanatory models for the original data; sometimes even without the development of any theoretical considerations.

A second source of confusion arises from the fact that the same model has different functions as a subject area evolves. The central Davisian model has, at various times, been a stage in the process of theory-building (Davis 1899), a primary teaching model (Johnson 1919), the source of validation of a hypothesis (Brown 1960) and the basis for extension and restructuring (Carson and Petley 1970).

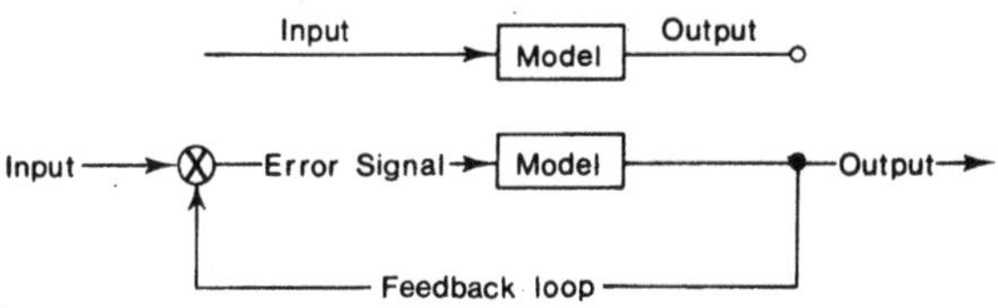

6.1 Diagram of a typical 'black-box' or transfer function approach to geomorphology in which an input is translated to an output over time and where a feedback loop operates at any given time within the model.

Chorley (1967) is responsible for a typology of the models used in geomorphology in what he calls a 'map' of geomorphic activity. This is based not on function but on 'the thought process of abstraction and decision making' which differentiates natural analogue, physical and general systems. The natural analogue systems are those in which an analogous system is 'believed to be simpler, better known or in some respect more readily observable than the original' and includes historical analogues and spatial analogues. The historical analogues lie closest to the qualitative temporal models described in this chapter. The other temporal models we have also distinguished in terms of philosophy of approach so the division is not identical with that used by Chorley. Nonetheless the general equation of physical systems with chapter 7 and general systems with chapter 8 may be a useful guide.

Harvey (1969) suggests an extremely useful set of 'procedural rules' which adequately qualified operate as a guide against which existing models may to some extent be evaluated. They are:

(i) the proposed function of the model must be specified adequately;
(ii) the function should not be changed without adequate safeguards;
(iii) models with more than one possible interpretation (e.g. regression models) or those for which theory is poorly developed should be avoided or improved;
(iv) a model should be identified with one and only one theory;

(v) conclusions drawn regarding a theory, from the manipulation of a model for that theory, should not be accepted automatically unless the model is identified with the theory or the domain of the model and its relationship to the theory can be specified fully;

(vi) where a model is used to predict, the conclusions drawn from that prediction are only fully acceptable when the model has adequate theoretical backing;

(vii) the multiplicity of functions, types and definitions given to the concept of a model should be appreciated in any research design.

Temporal qualitative models

Because paradigms change, the particular framework in which we operate determines the subject area, the kind of models we use and the solutions we adopt. The current paradigm is one in which process studies prevail effected principally and increasingly through mathematical and stochastic models. By and large, however, these models have developed from extensions of earlier models which were and remain essentially qualitative in character. Most of these have been subject to testing with some quantitative information and certainly quantitative information is involved where chronologies become fixed. In essence, though, the core of the argument has usually been expressed

Table 6.2 Examples for different qualitative geomorphological model formulations

		TIME STATUS		
		Floating equilibrium models (negative feedback)	*Relative time models* (positive feedback or cyclic)	*Fixed time models*
PROCESS STATUS	*Global* or *all process models*	Dynamic equilibrium model	Cycle of erosion model	Regional denudation chronology model e.g. south-east England
	Climatic, structural or *tectonic models*	Models of climatic control of process	Glacial-interglacial climatic sequence models	Regional climato-genetic complex model e.g. central Spain
	Single process models	Transport or weathering limited slope development models	1. Positive feedback glacial erosion models 2. Seasonal cycles of solute yield	Regional glacial, coastal, terrace or sea-level models

in non-quantitative terms. It has to be admitted that the subdivision of qualitative models is here based on different criteria to that of *all* temporal models, though the same criteria could also be used for deterministic and stochastic models.

Basically, qualitative temporal models lie along two continua (table 6.2). In the first, models have varying degrees of time fixedness. Some are independent of time, though by definition the expectations of these models can be observed only if time passes. Such 'floating' models assume the operation of processes independent of the particular time framework because, on the scale over which the situation is observed, the variables describing the state of the system remain essentially stationary. At the other extreme particular events, accounted for by the model, are fixed with respect to other fixed events and the conventional absolute time scale is used. In between are models in which the relative positions of events assume great importance but without reference to an absolute time scale. The other continuum ranges from the totality of processes operating or assumed operating in a single model, to the operation of a single process. Somewhere along this line processes are made more explicit by the assumption of climatic, structural or tectonic causation. Generally speaking the tendency in deductive building of temporal models has been to move from fixed to floating time and from global to process-specific models.

The idea of time required, but independent of a fixed time scale of events, is generally characteristic of models in which the process or assemblage of processes and the corresponding forms are (i) assumed to be poised in some condition of dynamic stability or equilibrium over the particular period of observation, or (ii) observed on a time scale so short that trends, which would imply relative position, are not observable or are unimportant. Generally speaking models of this type incorporate negative feedback assumptions to explain the absence of trend.

Specific admission of a directed change in the 'states' of the system, whether they be defined by the values of the variables or the geometric properties of the landforms, incorporate the ideas of positive feedback implying progressive departure from the initial condition. They assume a sufficiently long period for trends to be observed and detected and generally include, over the very long term, cyclic change or closed system decay. The Davisian model is here the classical example.

Fixed time models have sometimes arisen from an imposition of a relative time model onto observed events, and in this way the relative model, checked and sustained against the observed events in time, acquires the status of theory. In other cases the fixed time models serve as the inductive step towards theory. The first is exemplified by L. C. King's (1962) attempts to impose a particular model, as yet unsupported by adequate theory, on to the sequential development of forms in Great Britain, the second by the model of cyclical but overall decline in sea-level throughout the Quaternary adopted by Zeuner (1952).

Contemporary fashion for the moment encourages geomorphologists to operate mainly in the floating-time model-building framework. There are two good reasons for this. First, it is evident that our ability to fix events in time over the longer range and hence to test the relative and fixed point models is inadequate. Secondly, the assumptions of equilibrium of one kind or another facilitate model-building, perhaps unrealistically, because time-varying controls need not be specified. It has to be admitted that the comfort provided by negative feedback is gained at the expense of a better understanding of long term development of form.

In the discussion which follows the main types of qualitative models are discussed in the framework described above. The examples are intended to illustrate not only the nature of qualitative models but some of the important areas of investigation in geomorphology since the turn of the century.

Global models

Equilibrium model

The general propositions behind models of landform change through time range from those which admit a rapid transient response of a system to the environmental conditions followed by a long period of essentially unchanged form, to those in which change occurs continuously though at varying rates along a prescribed time-form trajectory.

Gilbert (1877) is generally credited with the development of the macroscopic model of landform evolution in which fluvial and hill-slope processes taken together provide a balance (equality of action) between the ratio of erosive action, as dependent on declivities, and the ratio of resistances, as dependent on structure. In this way any variation or disturbance to any part of the system would be passed throughout the river basin so that form becomes universally adjusted to process and that 'there is an interdependence throughout the system'. Hack (1960, 1965, 1966) has interpreted the Appalachian landscape in terms of the principle of dynamic equilibrium because of his inability to recognize peneplains or cyclic surfaces in the region. The model rests on the principle of dynamic equilibrium asserting that:

> When in equilibrium a landscape may be considered as part of an open system in steady state of balance, in which every slope and every form is adjusted to every other. Changes in topographic form take place as equilibrium conditions change, but no particular cycle or succession of changes occurs through which the forms inevitably evolve, as was assumed by Davis and most later workers in geomorphology. Difference in form from place to place are explained by differences in bedrock or in processes acting upon bedrock. Changes which take place through time are a consequence of diastrophic changes in the environment or of changes in the pattern and structure of the bedrock exposed as the erosion surface is lowered. (Hack, 1960)

Thus, Hack suggests that once a dynamic equilibrium form has been obtained it will for a 'certain span of time' remain essentially unchanged in character as long as the external conditions remain constant. For the Appalachians, Hack recognized a graded ridge and ravine topography but no relics of prior equilibrium conditions. Remnants of erosion surfaces were considered irrelevant to the interpretation of the landscape; this implies that the only history to be established is the mechanism and *time of attainment* of dynamic equilibrium, namely the initial rapid period of change. In so doing, the worker focuses interest on the relative-time aspects, rather than equilibrium aspects and concentrates on changes in the external controls and the relaxation times.

Some slope studies (Hutchinson 1967, Welch 1970, Brunsden and Kesel 1973) have tried to formalize the relative model and fix it in time, demonstrating for particular processes and particular lithologies an initial period of rapid adjustment followed by long periods of unchanged form. They claim that this is wholly explicable in terms of the properties of the slope-forming materials in a given set of environmental conditions. In a study of the diastrophic history and erosion surfaces of Africa, King (1955) and Pugh (1955) have argued that isostatic compensation consequent upon erosion of the surface material will take place revealing the existence of dynamic equilibrium. Recent work in New Guinea (Twidale 1968; Pain 1973) has suggested that such conditions are a reality. Here in a region of rapid uplift, tectonic instability and high rainfall, rapid erosion of debris leads to a uniform removal of debris from all parts of the slopes, parallel retreat and maintenance of a characteristic slope and ridge shape. This is a quantitative result in keeping with the principal of dynamic equilibrium.

However the difficulty of accepting this model as *the* scheme for geomorphological investigation seems to lie not so much in the presence or absence of erosion surfaces, but in the conceptual adjustment necessary with a change of time scales and a shift from independence in the variable system. There can be little doubt that river channels adjust many of the variables of their geometry to minimize the effects of changes in any one of them, but this in part reflects the very short period over which this type of observation has been made. The complex changes in climate also mean that 'external' controls have not remained the same and lead us to doubt not so much the validity of Hack's model in the short term, as its utility in the longer term. However, the model has proved singularly convenient to those who argue for the short term development equilibrium models by negative feedback. Moreover, it provided a lively and provocative antithesis to more established reviews. Finally, it drew attention away from the strange notion that time actually begets change and highlighted instead the relationships between change (or lack of it), process and form.

Decay and cyclic models

In equilibrium models, because the values assumed to be taken on by the form characteristics are allowed to oscillate only slightly about some mean position, and are always restored towards it by negative feedback, *then the point in time at which we observe the system is immaterial.* In decay and cyclic models, however, it is essentially assumed that the changes taking place are such that the system has a different configuration when observed at different times; in other words, landforms have an *observable* history.

These decay and cyclical models have been developed principally with a view to (i) understanding the evolution of contemporary forms by using a taxonomy based on existing features, (ii) extending the geological record (Wooldridge 1948) and (iii) using the models as a yardstick against which to evaluate the progress of a particular area through the evolutionary sequence defined by the model. The intellectual environments in which the more important of these models developed have recently been reviewed in detail and the reader is referred to the work of Stoddart (1966), Chorley (1965) and Chorley, Dunn and Beckinsale (1964, 1973). The last is the most comprehensive and readable account of W. M. Davis' life and work.

Davis considered landforms to be controlled by the interaction of three variables, structure, process and time, so that in a given climate and in relation to a given base level the landforms would develop through a sequence of stages from youth, through maturity to old age. A geographical cycle, said Davis, 'may be subdivided into parts of unequal duration, each one of which will be characterized by the degree and variety of relief, and by the rate of change that has been accomplished since the initiation of the cycle. There will be a brief youth of rapidly increasing relief, a maturity of strongest relief and greatest variety of form, a transition period of most rapidly yet slowly decreasing relief and an indefinitely long old age of faint relief in which further changes are exceedingly slow. Each one merges into its successor, yet each one is in the main characterized by features found at no other time'. (fig. 6.2a)

Davis went on to discuss the effects of *accidents*, both climatic (glacial and arid) and volcanic, and *interruptions* to the cycle where changes in base level by relative movements of land and sea led to renewed erosion and thus to polycyclic, partially complete, partially destroyed surfaces. The ideal cycle was presumed to lead by processes of hill-slope and relief decline towards a penultimate, baselevelled surface, the peneplain, in which all the slopes, waste mantles and streams were in a graded condition. Davis recognized that this particular model faced the difficulty of insufficient time for peneplanation, particularly in unstable areas, and recognized the need to allow for a multi-cyclic development to match the landscape as he saw it.

To many geomorphologists the Davisian model formed the essential basis that had been required to establish the life history of a landscape. The method was simple. First, if surfaces or partial surfaces could be formed then they could be recognized, mapped and placed in sequential order. Secondly, if an area had passed through several former cycles of erosion it would contain relics of dissected erosion surfaces which, by uplift, had formed the initial surface for the new cycle. By plotting these remnants and examining the accordant summit heights these surfaces could be reconstructed. The idea that an area could be considered to have been rejuvenated by uplift but still to bear traces of its former condition is still disputed today but has been used to establish many diastrophic and denudation histories. Thirdly, if dateable deposits could be found on the surfaces or if, by normal stratigraphical means, they could be related to known geological events, exhumed surfaces or dateable marine deposits, then the life history became more precise and comparison could be made with the relics of other areas. Fourthly, if the erosion surface record was poor or if deposits were absent recourse could be made to an examination of the drainage pattern.

The consequences of acceptance of the initial model, largely deduced in the fashion of the last paragraph, were embodied in the teaching and practice of denudation chronology. The central model has come under attack on four particularly important points:

(i) the inadequate theoretical understanding of the way in which processes lead to the sequence described by the model, for example that denudation is directly proportional to relief. In opposition, Penck (1924) and King (1957) presented alternative global relative-time models based on very different, though better defined, process mechanisms;
(ii) the tendency, at least on the part of those who followed Davis' teaching, to believe that the passage of time itself was somehow responsible for the succession of events;
(iii) the inadequacy of the model when faced with the evident mobility of the earth's crust, fluctuations in the climate and, amongst other things, fluctuations in sea level;
(iv) the problem of assuming that a particular form characteristic can be produced by one and only one mechanism. This issue is still unresolved in most geomorphological studies and, at least in the realm of model-testing, is no less true of contemporary mathematical models than it was of the Davisian model.

As Chorley (1965) has noted the main development of denudation chronology actually preceded the Davisian cyclic model. It was the apparent ease with which the chronological succession could be matched by features consequent upon the model, that was responsible for sustaining the model.

Whilst recent work has mainly challenged the inadequacy of the process assumptions, it has at the same time indicated that some of the assumptions in the model are quite reasonable (Schumm 1956, 1963a; Carson and Kirkby 1972). Others have attempted to restructure the model in the light of more coherent process arguments (Carson and Petley 1970).

Two particular developments caused mild disturbance amongst those who, in their chronological studies, followed the Davisian scheme. These were the models produced by Penck and King. The first questioned the tectonic assumptions implicit in the earlier model, and viewed landform development in terms of varying rates of tectonic uplift. Penck's model also involved the *primärrumpf* which represents the extreme case of the equifinality problem raised in (iv) above. This was the form produced when a surface underwent slow uplift, exactly matched by denudation, so that there was no increase in net relief!

The second worker, L. C. King, questioned the process assumptions made by Davis whilst accepting the idea that 'following the creation of new surface states by tectonic movement', the landscape will be degraded by surface processes. King, however, appears to have founded his work on Davis' mistranslation of Penck's ideas of slope development, the descriptions of slope form of Wood (1942), the *piedmonttreppen* concept of Penck and a combination of both the multiconvex *primärrumpf* and the multiconcave

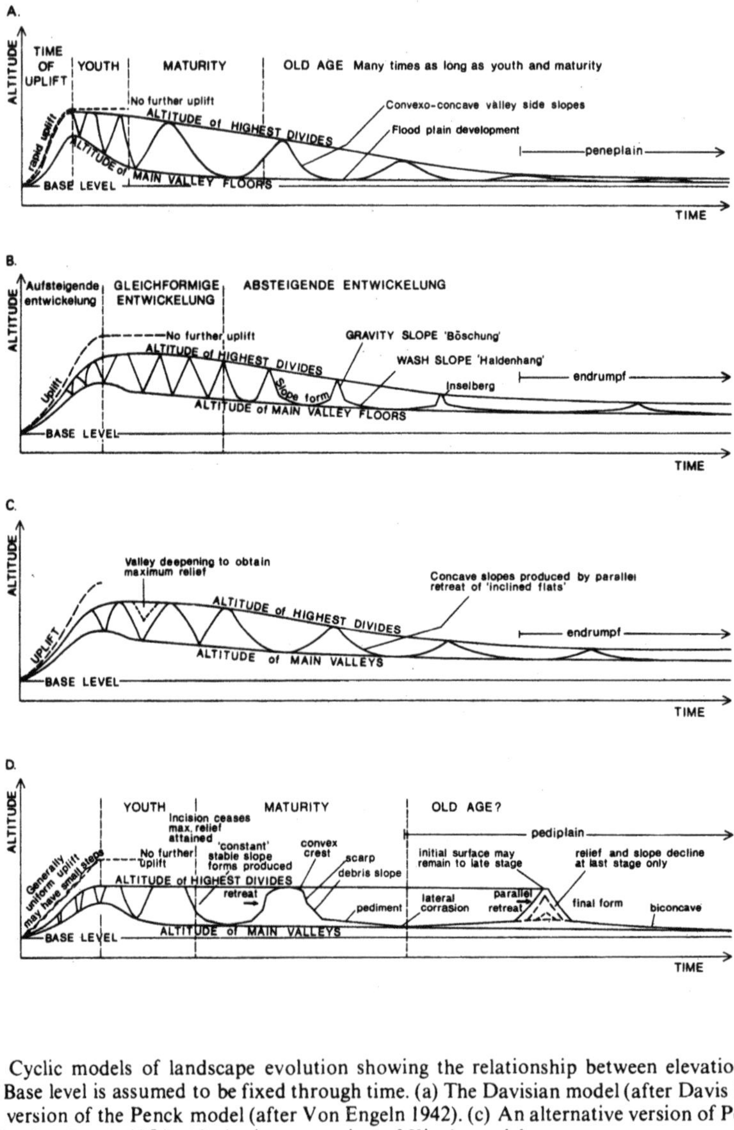

6.2 Cyclic models of landscape evolution showing the relationship between elevation and time. Base level is assumed to be fixed through time. (a) The Davisian model (after Davis 1899). (b) A version of the Penck model (after Von Engeln 1942). (c) An alternative version of Penck's model (after Penck 1954). (d) An interpretation of King's model.

endrumpf which are accepted in terms of rock strength rather than rate of uplift. To all this he adds a hillslope model based on parallel retreat and leading to the formation of a pediment (Bryan 1940).

The important consequence of both these works for denudation chronology was that the temporal relationships derived on application of the King or Penck models were very different from those derived from Davis' model (fig. 6.2a,b,c). Surfaces could be of metachronous (time transgressive) character; scarps could remain youthful until a late stage in the cycle but were to be part of the same cycle as the old-age form; successive cycles of different age could all be denuding the landscape at the same time, so that the upper parts of different surfaces could be the same age.

Fixed scale global models

One of the important features of the Davisian model, derived from the earlier work of Powell (1875), was that erosion was controlled by available relief as measured above a base level. The level of the sea was imagined as extended through the continents, though the earlier usage was different. In the ideal cycle of erosion model, this base level was assumed fixed throughout the cycle (fig. 6.2a) though Davis allowed an alternative model in which the progress of the cycle was interrupted; so that only partial fulfilment of the cycle occurred. Using these ideas denudation chronologists sought to relate their field observations to known or inferred changes in the base level resulting from either uplift or eustatic change, to develop new models for historical development called regional chronologies.

One such regional model was provided by Wooldridge and Linton (1955) for south-east England. Following the events of the middle and late Tertiary, they induced a series of falling base levels interjected by longer periods of sea-level stability during which partial cycles were developed along the main river channels in the area. The evidence for the model comes from the character of the drainage pattern, the nature and distribution of erosion surfaces, deposits whose age is known or may reasonably be inferred, and the evidence of eustatic changes in sea-level elsewhere, notably in the Mediterranean. Prior to these events, the chronological model specifies the tectonic and denudation events of the early and mid-Tertiary. The basic features of the model are shown in table 1.2 fig. 6.3. This model served as a standard for chronological research in Britain and has only recently been re-evaluated in the area of its original conception (Jones 1974). The model was later applied over widely-ranging rock types and physiographic conditions; in areas where warping is known to be active, even at the current time; and in areas where intense glaciation might have been expected to remove all traces of the partial cycles. It has been attacked on grounds of inadequate evidence of process (some believe the surfaces to be of marine origin, for example); inadequacy of dating (notably the continuing debate about the age and origin of key deposits) and of problems in some critical areas (notably the drainage pattern). Finally, the model has been extended in domain to continental Europe and even, by comparison, to North America (Brown 1961).

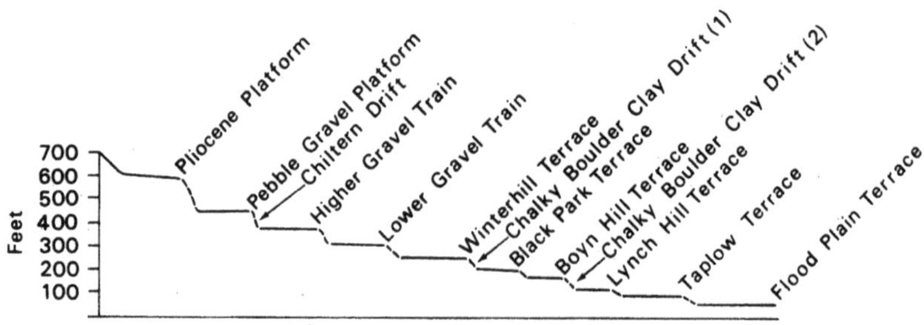

6.3 Part of the regional model for the London Basin developed by Wooldridge and Linton (1955). The model shows terraces produced by an assumed falling sea level with periods of sea-level stability.

Process-assemblage models

Floating chronology

Unlike the global models, which have essentially been process independent, most models are to some extent specified. That is to say, unlike the global models, they are assumed to operate under spatial or temporal constraints depending on whether a particular set of circumstances exist. Climatic assemblages have been singled out for treatment here but other qualitative models have been based on tectonic zones, lithological units ('Karst' models) or particular environments (coastal or fluvial). The models based on premises about climatic control of landforms have been especially important in relation to the development of sequential and fixed point models and will thus serve as an example here.

The equilibrium models of climatic-process control were developed principally in Europe in the latter part of the last century and seem to have developed in parallel and in a somewhat confused fashion along with relative and fixed-time chronologies. The arguments were stressed by Peltier (1950) and most coherently evaluated by Stoddart (1969). A recent reader provides translations of the most important papers (Derbyshire 1973).

In essence the argument runs as follows:

(i) A particular region has a given climate defined in terms of parameters such as mean annual precipitation, mean annual temperature, distribution of rain days, frequency and magnitude of precipitation events and so on.

(ii) The particular processes in operation, and more especially their rules of operation, are essentially determined by climate and so may be characterized in terms of particular climatic parameters.

(iii) A particular assemblage of processes gives rise to a particular, identifiable set of landforms which therefore relate to a particular climate.

(iv) As in the global equilibrium model, landforms are assumed to adjust rapidly (over a relaxation period) to any change in the climatic parameters and thereafter to remain in equilibrium with the existing parameters.

The shortcomings of this structure lie in the failure to test the four assumptions, though intuitively we would question the utility of such simple parameters as mean annual precipitation or temperature. This is a matter of sophistication and a deeper understanding of climate-process relationships; the basic relationships are being actively examined under the current research paradigm.

Much more serious is the uncritical acceptance of this overall model and its application in three areas: (i) as an explanatory model for present-day forms which appear at variance with present-day processes, by the insistence that the present landscape can only be understood in terms of past climates, (ii) as a model for explaining the past sequence of landforms, by invoking a set of past climates, and (iii) as a model for inferring past climates by assuming the origins of past landforms. The facility with which tautological statements appear in the literature is a measure of the confusion of these last three points in the minds of many authors.

Relative models

One group of relative models relating to climatic control is that in which the landscape evolves under a *given* climate through a fixed set of stages of evolution. These are directly comparable to the cyclic global models. Davis considered that the results of 'climatic accidents' would be modelled in this way and himself described cycles of erosion for arid and glacial conditions (1905, 1906) and the idea has found favour in several quarters (Peltier 1950 for a periglacial model; Birot 1960 for a general thesis relating to this type of model).

Another, perhaps more important, group of models arises from the deduced effects of an assumed series of changes, such as those taking place in the Pleistocene. A great deal of work has been done in this area, much of it highly speculative and poorly based. Nearly all of it faces the basic difficulty that the very features which we seek to explain are those providing the evidence of climatic change in the first place. This is particularly so where information on climatic change from alternative sources, such as pollen analysis, is not available. Moreover some would argue that 'The pervasiveness, intensity and numbers of Pleistocene climatic changes precludes the widespread development of truly climax (dynamic equilibrium) landforms within most existing floral morphogenetic areas' (Garner 1968).

An important model for determining the expected response of river channels and hillslopes to changes in climate was suggested by Schumm

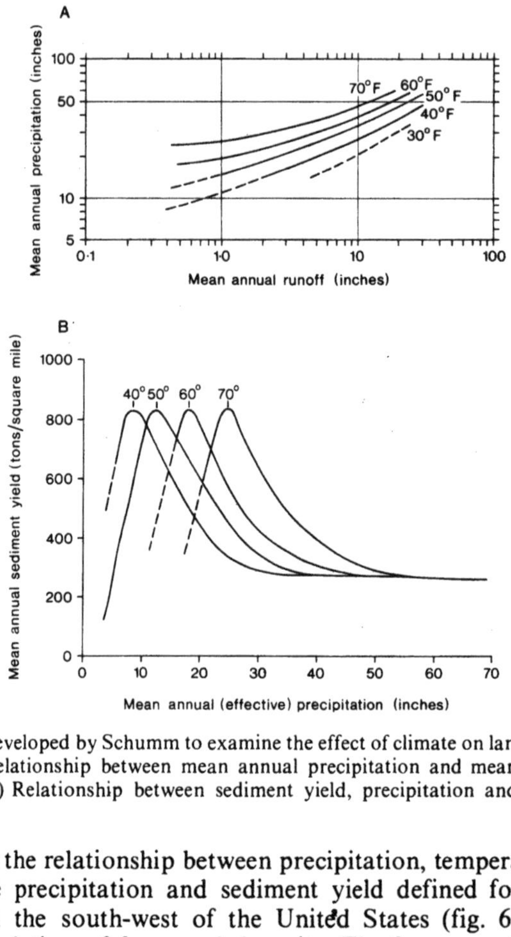

6.4 Graphs developed by Schumm to examine the effect of climate on landform and sediment yield. (a) The relationship between mean annual precipitation and mean annual runoff and temperature. (b) Relationship between sediment yield, precipitation and temperature (after Schumm 1965).

(1965), using the relationship between precipitation, temperature and runoff and effective precipitation and sediment yield defined for data collected mainly from the south-west of the United States (fig. 6.4a,b) and long distance correlation of form and deposits. The few examples that do exist seem to be better related to specific processes than data hitherto used.

Fixed models

Fixed-time models of geomorphic changes in climate and form are almost as numerous as the regions whose changes they seek to portray. These models are typically based on three types of evidence: (i) dateable deposits separating undated deposits, (ii) non-sequences and unconformable relationships, and (iii) form relationships. Most typically the deposits comprise relict soils and various types of crust, peat or other pollen-bearing organic sediments and a variety of terrace, marine or colluvial deposits which are usually credited with definite and distinctive climatic implications. Form relationships include the height of terraces, lake shorelines and marine beaches, drainage density and characteristics and the juxtaposition of various sedimentary bodies.

Fixed chronology models of fluvial terraces of Quaternary origin achieved great importance particularly in the earlier part of the century and the ubiquity of rivers having four terraces is quite astonishing, despite the fact that Soergel (1924) for example had demonstrated the existence of not less than ten aggradation stages along the river Ilm in Thuringia. Later studies have proceeded with much greater caution, recognizing the complexity of possible responses by basins to climatic changes as well as the added complication of Quaternary and recent tectonics. Just as the earlier models laid most of the changes at the doorstep of eustatic changes later models of terrace form and morphology have overstressed the necessity for a direct relationship with northern latitude glacial chronology. The dangers and problems of such assumptions have been highlighted in the Quaternary stratigraphy of the American south-west. Here the simple equation of pluvial-glacial has for long been in considerable doubt. The possibility exists, for example, that in the Mediterranean (Butzer 1964) there were probably two kinds of 'wet' period, cold-wet pluvials with a strong westerly regime (in early and late glacial times) and a warm-wet pluvial representing the warmest interglacial with arid phases in-between.

This type of work has been particularly important in these areas of the world where the more positive (but nonetheless difficult) evidence from glacial activity is lacking. Even in the periglacial areas the doubts about origin and environmental conditions made dating by reference to climatic change, or assertions about climate in terms of deposits a very hazardous business indeed.

The difficulties of this type of approach are exemplified by the work of Butzer (1964) who has been especially responsible for the elaboration of regional chronologies in the Mediterranean and in the Near East. Butzer examined later Palaeolithic sites at Torralba and Ambrona, south-east of Soria in Spain (fig. 6.5). The total section of several metres thickness was subdivided on the basis of sediment characteristics. These sediments are then given climatic interpretations by reference to analogues elsewhere in Spain. In turn, the climatic interpretations are used to suggest a dâting sequence tied into the faunal remains of the site, and with some palynological confirmation of the geomorphological evidence. Insofar as the archaeological and pollen evidence supports the 'geomorphological' evidence and roughly dateable material is available, the excavation is one of the more helpful and meaningful cases of climatic interpretation of stratigraphic evidence. For the most part this kind of geomorphological reconstruction when taken alone is singularly dangerous because (i) the process-climate relationships of even contemporary deposits are very poorly understood; (ii) the use of 'contemporary' analogues assumes that the analogues themselves are in phase with their contemporary environments; (iii) different processes can produce the same deposits, and (iv) sediments are subject to strong seasonality in processes of production and sedimentation. Often the rhythmic deposits called *éboulis ordonnés*, for example, can be seen forming today in areas of strong seasonal contrasts and slopes above 26° in the Mediterranean.

UNIT	SEDIMENT	THICKNESS (Cm.) T.	A.	ASSOCIATED PROCESSES	CLIMATIC INFERENCES	TENTATIVE CORRELATION
UPPER COMPLEX	IV Reddish colluvium	120	200	Soil colluviation with local fan alluviation at valley margins.	(Human interference)	Historical
				EROSION		
	III Fine dark alluvium	80		Valley alluviation	Moist, temperate	Middle Holocene
	II Coarse brown alluvium	70		Alluviation at valley margins by lateral tributaries; tufa deposits locally. Some solifluction initially.	Cool, moist	Würm Glacial
	I Reddish colluvium	125				
				EROSION		
MIDDLE COMPLEX	II Yellowish sands	10		Colluviation and valley alluviation, following intensive frost weathering. Slumping of subsurface, lubricated Keuper silts producing faulting at both sites. Some solifluction.	Cold, moist	Riss Glacial Complex
	I d. Reddish colluvium	55				
	I c. Reddish alluvium	60				
	I b. Reddish colluvium	30				
	I a. Cryoclastic detritus	20				
				EROSION		
PEDO-GENESIS	Terra fusca soil developed on Lower Complex IV and V exclusively. B	160	150		Warm seasonally very moist	Great Holsteinian Interglacial (=Tyrrhenian I stage)
	Bc	10	35			
	Ca	10-20	10-60			
				EROSION		
LOWER COMPLEX	V d. Coarse reddish alluvium	165	95	Shallow* alluviation at valley margins by lateral tributaries.	Very cold	Stadial (LATE ELSTER)
	V c. Fine reddish alluvium		85			
	V b. C-gravel		60			
	V a. Gritty marls		90			
						Interstadial
	IV b. Gray marl	150	200	Valley back swamps filled with homogenous fine silts from sluggish flood waters; pseudo-gley conditions indicated by limonitic Fe-horizons.	Moist, temperate	Stadial
	IV a. Marl with channel beds		220	Valley flood-plaining dominated by very fine alluviation, but with coarse, moderate cryoclastic channel beds locally.	Moist, cool	
				EROSION		
	III b. B-gravel	–	15	Coarse valley alluviation with reduced soil frost.	Moist, cool	
	III a. Upper gray colluvium	80	80	Fine valley alluviation with some solifluction.	Cold, moist	
				EROSION & CONGELIFLUCTION		Interstadial
	II d. Sandy marl	90	150	Fine valley filling.	Moist, temperate	Stadial
	II c. Lower gray colluvium	100	?	Well-stratified gritty sands with intercalated gravels. Valley alluviation.	Cold, moist	
	II b. A-gravel	30	60	Coarse cryoclastic gravels on slopes, partly calcrete, partly interbedded with grey silts. Some congeliflucton.	Cold, moist	
	II a. Light sand	70+	300	Fine valley filling of homogeneous sands partly silty at top.	Cool, moist	
				EROSION		Interstadial?
	I Red colluvium	400+	?40	Medium, highly cryoclastic detritus at base of slopes.	Very cold	Stadial
	0 Redeposited Keuper (several phases)	100+	200+	Congelifluction and earth flows of lubricated clays, silts and marls.	Moist, cold	

6.5 The record of sediment, processes and climatic inferences discussed by Butzer (1964) for the regional chronologies of Torralba and Ambrona, southern Spain.

Qualitative models of particular processes

Equilibrium models

Insofar as equilibrium models are among the easiest to quantify, most of the models of this type have undergone further development. Typically they are characterized by negative feedback relationships which make them trend-free on the types of time scale over which they are usually observed, so that the geometry of the system is fairly fixed.

Gilbert's (1909) continuity model for hillslope form has this character and Gilbert uses it to argue for a fixed equilibrium geometry of convex shape. For any cross-sectional area the amount of soil material to be evacuated increases with distance from the divide. If constant depth of soil cover is to be maintained and weathering is assumed to be constant over the whole slope, to evacuate more material Gilbert required that the rate of movement increased downslope and this was provided for by increased slope with distance from the divide. Further downslope this mechanism was masked by the effect of running water under the action of overland flow. It is not, however, made clear from Gilbert's model why he should assume uniform thickness, especially in view of his earlier understanding of the importance of the balance between production and removal of material. The latter implies that rate of production of material is conditioned by the existing depth of soil cover; as the depth increases the weathering interface is 'cut-off' and weathering increases as soil thickness decreases, being at a maximum when soil cover thickness is zero. It also carried the implication that when there is no cover the development of a slope is limited by the rate of weathering; otherwise it is limited by the capacity for transport. Several of these features have recently been incorporated into mathematical models (Ahnert 1970, Kirkby 1971, see chapter 8).

Relative models

It is perhaps in the area of relative modelling of contemporary processes that qualitative, indeed any kind of modelling, has been extremely active. Unconstrained by the feeling that the models needed to be specified closely, by a scientific conscience that insists on rigorous testing or indeed by the probability that their models would ever be tested, geomorphologists built and continue to build a large number of such models.

Many of these models are specific to a particular area or even phenomenon, a situation which stems from the inductive, genetic philosophy of the first half of the century. Others are universal, well formulated and capable of rigorous testing.

The general models are of two types, positive feedback models (evolutionary) and cyclical.

We illustrate the first type with reference to C. A. M. King's (1969) positive feedback model for glacial erosion, based on the Nye (1952) model of glacier flow and applied to the Austerdal valley in Norway. Nye's model relates changes in the longitudinal strain rate of a glacier to variations in bedform

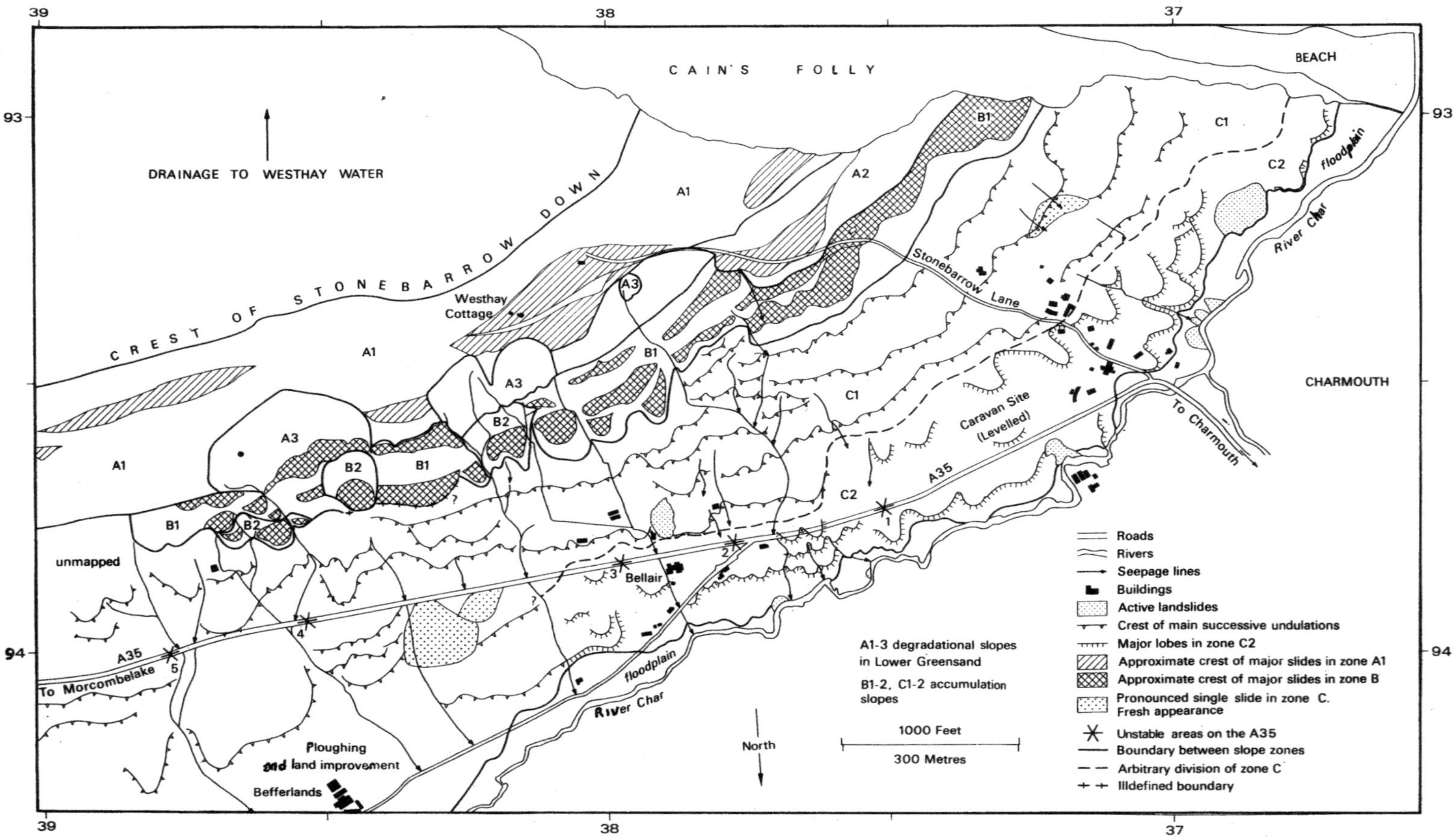

6.6 Detailed map of a landslide complex at Stonebarrow Hill, Dorset, the events of which were suggested using relative dating techniques (after Brunsden and Jones 1972). For full explanation see text.

and accumulation and ablation down glacier. The applicability of the Nye model is confirmed by field measurements of strain rate. The model is extended to yield greater erosion in the areas of bed concavity, tending to increase this concavity further and in turn strengthening the compressive flow. Added to this is the effect of ice thickness which according to King would produce more erosion by virtue of the greater amount of melt water and its effect on the rate of sliding. Thus the model envisages the progressive enlargement of pre-glacial changes in gradient to form the classic riegel and riser model of the glaciated valley.

Inductive development of a qualitative model is exemplified in some work on Dorset mudflows by Brunsden and Jones (1972). The authors describe in detail a large landslide complex by detailed mapping and classification of forms (fig. 6.6) which they attempt to date by relative means. From this morphological evidence they conclude that the following evolution of slope forms occurred (fig. 6.6):

1. Widespread landslide activity over a long period of time following the incision of the river Char.
2. A period of solifluction which led to extensive convex-concave slopes. In some areas landsliding may have continued during the phase of solifluctional activity (fig. 6.6 zone A1).
3. The erosion of most of the head-mantled slopes by widespread large-scale landslide activity (fig. 6.6 zones A2, B1).
4. The continuation of landsliding in wet areas leading to the formation of new, lower landslide scars (fig. 6.6 zone A3).
5. Localized activity diminishing with time. The degradation of the landslide units and subsequently the present movements.

The authors then go on to suggest the possibility of a more general model in which as a slipped mass moves forward, a further landslide occurs in the degradational zone above so that a sequence of features develops through time which is also exhibiting progressive degradation in space.

Evolutionary models have also been popular and important in glacial geomorphology. Here an example by Price (1969) is very similar in approach to the work just described. From the analysis of a highly complex pro-glacial environment Price induces a more general model which may have widespread application and may be expressed by an idealized block diagram (fig. 6.7).

Cyclical models for individual processes tend to relate to either (i) processes related to seasonal climatic phenomena or (ii) deductions as to the long-term performance of a system. Edwards and Thornes (1973) have made a study of the cyclical character of dissolved solute concentrations in the river Stour of Essex and find marked annual peaks in most of the measured constituents. These constituents are staggered in time and a model in the form of an annual 'clock' describes the succession of solutes in relation to the annual march of climatic and edaphic conditions. Again the model is essentially inductive.

Deductive cyclical models for specific processes have tended to follow the lines set by the Davisian model in being highly speculative, often inade-

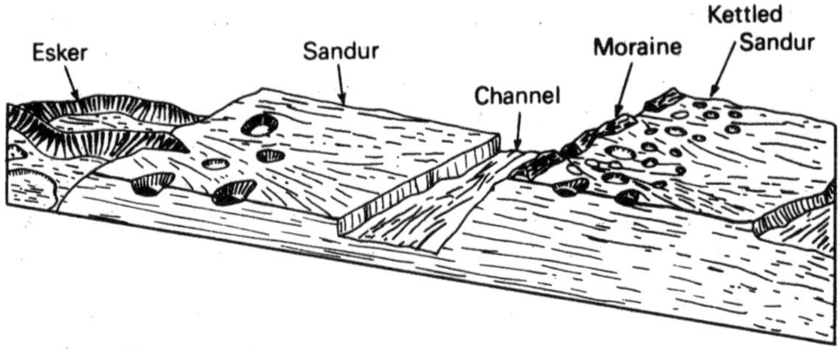

6.7 General model suggested by Price 1969 for a complex pro-glacial environment. After a study of specific glaciers in Iceland, Price established the sequence of landforms shown above which he suggested might be regarded as a model suitable for the study of glaciers in other areas.

quately based in either theory or observation and sometimes quite incapable of testing. Such models include Davis's (1905, 1930, 1933, 1936, 1938) own cyclical model of arid landform development, Peltier's (1950) suggested scheme for periglacial process and form relationships and Johnson's (1919) model for coastal evolution.

Fixed models

For the sake of completeness we mention some of the best known, fixed chronology, qualitative models for the response to a process in a particular place or environment. The models are essentially historical analogues in the terminology of Chorley (1964). They include the well-known models of regional glacial chronologies; chronological models for coastal evolution of specific areas (such as the mouth of the Mississippi) or for terrace sequences (such as the Thames); and the global chronologies for sea-level change.

These models exhibit characteristically high levels of abstraction both spatially and temporally as well as in details of process. They have the primary function of simplifying the sequence of events, firstly in order to 'explain' the existence of particular features extant on the earth's surface at particular locations and, secondly, to compare with studies from other areas. This comparison is designed ultimately to produce a wider theory of landform genesis and in some fields this has been the case. Local studies produced models which now yield a coherent theory of glacio-isostatic recovery; perhaps one of the best examples of the possibilities of this kind of model-building. Progress has also been made along these lines in the development of glacial theory, in the study of regional tectonic activity and in the understanding of global sea-level change.

We have exhibited here but a few very varied examples of qualitative model-building. It has and continues to be a very fruitful area of geomorphological endeavour. In the succeeding chapters, we examine the more recent developments in model-building using mathematical and statistical techniques.

7 Quantitative deterministic models of temporal change

In an earlier chapter (3) we introduced the notion that events occurring in time could be regarded as having varying degrees of 'memory'. Models in which events in time are completely independent of all previous events were said to exhibit complete randomness; very few situations in nature really have this character. A continuous random series, while being useful as a concept in an abstract sense, is virtually unobtainable. At the other extreme all events are entirely prescribed and the system under observation is assumed to have an infinitely long memory. These are the deterministic models of the type to be examined in this chapter and the methods used for their examination, development and testing are those of applied mathematics, especially the differential calculus. Differential equations, with all their ramifications and generalizations, are undoubtedly the most powerful tool in applied mathematics and it is hardly surprising therefore that the models used take on this form.

It is useful at this point to recall the reasons for adopting deterministic mathematical models relating process to response, input to output, cause to effect. The first is that provided our conception of the process under study can be transformed into the language of mathematics, then there exists a whole system of techniques and a body of theory for manipulating the relationships in the model in an objective and replicable manner. In other words, the procedure offers great facility provided that we are familiar with the language and technology of applied mathematics. This procedure of transforming a problem, performing some kind of operation on it and then reversing the transformation is the very essence of mathematical technique, as well as the reason for its adoption in geomorphological research.

The great progress of classical physics in the first part of the nineteenth century, stemming from Newton's law of gravitation and fully developed by Laplace and Lagrange, depended entirely on simplification and abstraction and rested in the belief that the universe was rationally constructed. The fundamental proposition, that abstract models constrained only slightly by the limitations of experience were of the essence, still holds today even in those sciences such as geomorphology where mathematics has been relatively recently applied.

A second major reason for adopting an applied mathematical approach to process-response modelling is the adoption of a systems approach in the subject at large. While there is still a widespread and elementary view, to some extent propagated in recent literature, that the 'systems approach' consists rather largely of 'organizing' things in boxes or of making broad and not very useful statements about interactions, quite the converse is true. This attitude seems to stem largely from the difficulties of accepting the engineering and applied mathematical techniques in areas, not traditionally mathematically-based, such as sociology and human geography. Unhappily, geomorphology still lies in part, at least, in this camp, though books such as Chorley and Kennedy's *Physical Geography. A Systems Approach* have gone some way to improving the situation.

Given a simple input or cause or process in a system, this may be transformed by a transfer function into an output effect or response. Mathematically, the most simple system could take the form $Y_t = C. X_t$ where X_t and Y_t are the input and output respectively at time t, C the transfer function. In a most general fashion, three typical problems arise: (1) given the input and transfer function determine the output, (2) given the output and transfer function find the input and (3) given the input and output find the transfer function. The systems are usually described in terms of differential equations which almost invariably involve a time element and, of course, the transfer function embodies the characteristics of the system. The response of the system to various kinds of input is determined by the nature of the differential equations which describe it and systems are classified in terms of the order of these equations and the typical temporal response they produce. The ultimate goal is to obtain the laws which define the system (and hence the transfer function) from completely theoretical assumptions, so that the

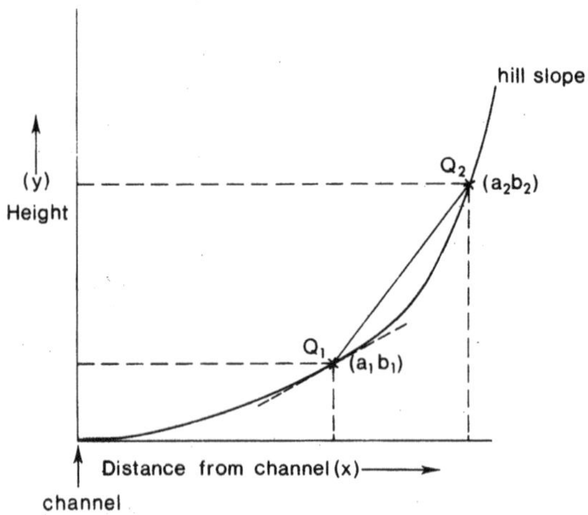

7.1 Graph to show the derivation for a tangent to a curve.

output is defined for any input. Many attempts to define transfer functions in the natural sciences have been empirical rather than theoretical in nature; some of these are dealt with in the next chapter. It is often argued that they are more concerned with prediction than understanding. At this point we simply wish to stress that deterministic and stochastic models meet on common ground in systems analysis.

The third major reason for adopting a deterministic quantitative approach to process response modelling is that between them Newton and Liebnitz provided a special set of techniques for dealing with rates of change, the differential calculus. Newton's three laws of motion involve the idea of speed and rates of change, whereas Liebnitz was concerned with the formula for the gradient of the tangent to a curve. In most geomorphological applications both approaches are used interchangeably, the change of height with distance is directly analogous to the Liebnitz formulation since the ground slope is a tangent to this curve. The decrease in height of a point over time is directly Newtonian. Before using these ideas further to develop differential equations of temporal change, a brief digression into the symbols and terminology is necessary by way of revision.

A digression into elementary calculus

A basic notion in calculus is the function: a set of ordered pairs such that no two ordered pairs have the same first element. Most frequently, these order pairs are an independent variable x and a dependent variable y. This idea is expressed as $y = f(x)$ where $f(x)$ formally means the value of y when x takes on a particular value. This is often generalized so that $f(x)$ refers to any equation in the variable x, and we say 'y is a function of x'. The value of y may be determined by two or more variables, e.g. z and x, then $y = f(z, x)$. Simple functions can be illustrated graphically (fig. 7.1). Assume we have a hillslope and we observe two points far apart. For each point we could observe the height and distance from the channel at the foot of the slope. The gradient of the line is given by $b_2 - b_1/a_2 - a_1$ for the point Q_1, where a and b are the Cartesian coordinates of the two points. Obviously, the closer Q_2 approaches Q_1 the better the estimate of the slope at Q_1. As Q_2 approaches Q_1, instead of being a chord, it reaches the point at which it is tangent to the curve at Q_1. This is the limiting position. It would then be meaningful to define ground slope by the tangent to that point on the curve. This procedure is empirical: we could go out and perform it in the field. Liebnitz however assumed that the slope could be defined by a function and found a method for deriving the tangent to any point on the curve. Given that height (y) is a function of distance from the channel (x) how can we derive the tangent to the slope at any point, which is the change of height for a small change in x at any value of x? This process is differentiation. In the Liebnitz notation, this tangent to the curve is the derivative and is expressed by the notation dy/dx or $f'(x)$. The expression d/dx is simply an operation for transforming a function into its derivative. If f' itself is a function, it too can be differentiated to obtain f'', d^2y/dx^2 or the second derivative. The first derivative of the function describing the relationship between height and distance is a function describing slope at any point. If this is further differentiated we have the change of slope with distance, which is of course curvature. The mathematicians have developed a set of rules for differentiation and a good introduction is given in the inexpensive text by Hilton (1968). In many slope models relatively simple relationships are assumed between height and distance, so that very difficult differential equations are avoided. Natural slopes are quite complex, but by using Taylor's theorem, the derivatives may be obtained for some polynomial of a high order (in other words a complicated function of height). Conversely, by knowing the values of the derivatives of f (x) of various orders at a set of points, we could reconstruct the polynomial which describes the slope, and hence if we wished to use it to predict the slope over distances. Such a procedure relies on the fact that the function is 'smooth' i.e. has no relatively sharp breaks in it; this assumption is often made for convenience in mathematical slope modelling.

We have used height and distance to review the notion of the derivative, since most geomorphologists are familiar with them. The extension of first and second derivatives of the height and distance matrix into three dimensions has been especially considered by Evans (1972).

In three dimensions, the height z can be described as a function of map co-ordinates x and y, so the function can be written as $z = f(x,y)$. In this case the process of differentiation has to take into account the fact that a small variation in z cannot be a function of x alone (except in cross-section) or y alone (except in another cross-section at right angles to the first); z is partially dependent on x and partially on y. This is also true of the derivative and the procedure for differentiation is known as partial differentiation; z is differentiated with respect to x whilst y is held constant. The curled ∂ is used so that the result will clearly be distinguished from ordinary differentiation, so that the two cases are represented by $\partial z/\partial x$ and $\partial z/\partial y$ respectively. Geometrically (fig. 7.2) $\partial z/\partial x$ respects the slope of the curve cut from the surface $z = f(x,y)$ by the plane $y =$ constant. If both these are allowed to vary, then the total differential would represent a change in the z co-ordinate of the tangent plane to the surface. As with ordinary derivatives, so with partials we

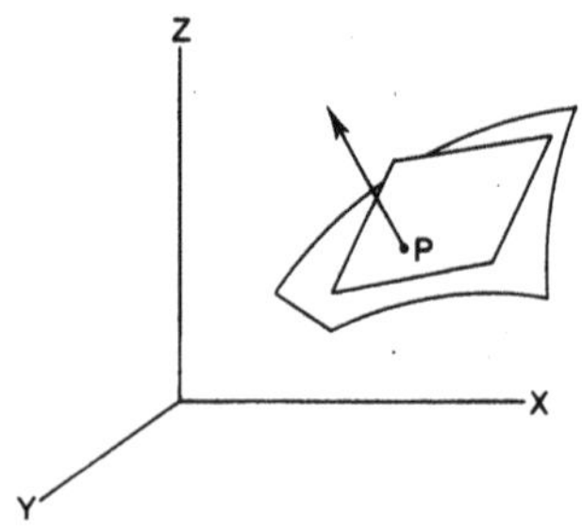

7.2 Tangential plane to a surface at a point P. The arrow indicates the normal to the plane.

may obtain second, third etc order derivatives represented by $\partial^2 z/\partial x^2$ etc. Equally, there exists a set of techniques for obtaining the partial derivatives, of any order. Finally, suppose that a variable z is a function of two other variables and these in turn are themselves functions of another variable, such as time, then z is a function of t and may be differentiated with respect to t, then

$$\frac{dz}{dt} = \frac{\partial z}{\partial x} \cdot \frac{dx}{dt} + \frac{\partial z}{\partial y} \cdot \frac{dy}{dt} \tag{7.1}$$

For example, let z be the height of a point on a former strandline, x be the rate of isostatic recovery and y the rate of sea-level change; then we could have $y = f(t)$ and with the derivatives dy/dt *and* dx/dt, each the function of a single variable. We have two further partial derivatives if $x = f(x,y)$ and so we have a general expression for dx/dt.

It is no surprise to find that the expression for the ordinary derivative occurs relatively rarely in geomorphic model-building. Invariably, we are examining situations in which some variables are held constant, not least of which is usually the horizontal spatial co-ordinate, for example in slope studies.

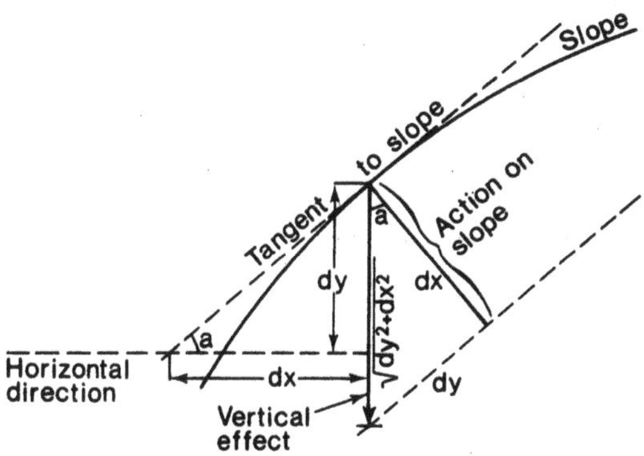

7.3 Terminology for the derivation of slope evolution equations (after Scheidegger 1971).

Differential equations

Consider a hillslope in which denudation is proportional to the height of a point under consideration above a certain base level. Scheidegger (1961) suggests this elementary model to introduce more complex models (given the assumption that precipitation increases with height, this is not absurd). The height loss considered will be measured vertically (fig. 7.3), so that the model can be expressed by saying

$$\frac{\partial y}{\partial t} = -y \qquad y = \text{height}, t = \text{time} \tag{7.2}$$

This is a differential equation of the first order because the equation contains a derivative and the derivative is the first derivative of some function. A solution of a differential equation is that expression for the dependent variable which does not involve any of its derivatives and which, when substituted into the given equation, reduces it to an identity. The solution to this particular differential equation is given by

$$y = f_0(x)\, e^{-t} \tag{7.3}$$

where $y_0 = f_0(x)$ describes the original land surface which is subject to change. For example, suppose $y = 2x$ describes the original landform then the solution is given by

$$y = 2x \,.\, e^{-t} \tag{7.4}$$

which can be evaluated for given values of x after time t. If we further assume that some constant relates height and rate of removal, we shall have $y = 2x \,.\, e^{-ct}$ where c is the constant.

Reviewing this example, we note the following steps:

1. conversion of a verbal statement into a differential equation;
2. solution of the differential equation to obtain a derivative-free function of a most general character;
3. specification of the initial conditions;
4. substitution of any real parameter values (e.g. a value for the constant, c, into the solution, and
5. examination of the results.

It is quite important, in reading literature on quantitative deterministic models, to be able to identify the separate steps. If this is done, the reader will put himself in a more usefully critical frame of mind. Even the non-mathematical reader will be able to evaluate for himself steps 1, 3, 4 and 5, provided that he can sort out the steps. Differential equations and their solutions owe their importance to the fact that there is a clear correspondence between them and the situation they represent. They usually provide a clear and simply expressed model of a somewhat complex physical situation. Step 2 usually presents most difficulty. Most of the important differential equations of mathematical physics have been derived from the process of separation of variables (see p. 152 below). Their solutions have been given special names such as Bessel, Legendre and Mathieu functions and their properties described in reference books of mathematical functions. There are about 2,000 functions with known solutions in all. One consequence of this is the tendency of some workers to cast the problem and its differential equations in a form for which the solutions have been developed elsewhere. In three examples later, we shall show how this applies to slope studies and the diffusion equations, glacier flow and characteristics, and rejuvenation and perturbation theory.

The differential equation described above is of the simplest type. A somewhat more complex model is given by

$$\partial y/\partial t = a \cdot \partial^2 y/\partial x^2 \tag{7.5}$$

which is a second order differential equation (Culling 1960) in which the change of height (y) with time (t) is a function of the local curvature multiplied by a constant (a). The solution to this equation is known from heat-diffusion problems in physics and is given by and shown in fig. 7.4.

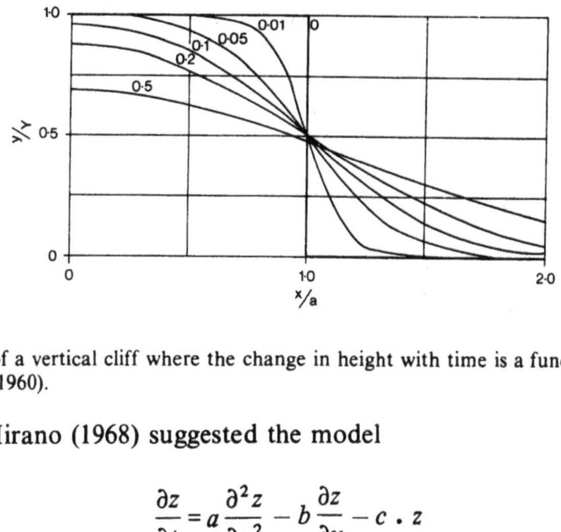

7.4 Decay of a vertical cliff where the change in height with time is a function of curvature (after Culling 1960).

Finally, Hirano (1968) suggested the model

$$\frac{\partial z}{\partial t} = a\frac{\partial^2 z}{\partial x^2} - b\frac{\partial z}{\partial x} - c \cdot z \tag{7.6}$$

(where a, b and c are 'erosional constants') which comprises a combination of the earlier models. The procedure of finding a solution is as outlined above. Once a general solution is found, then the parameters may be changed; again the differential equation is of second order. The other character which the three models have in common is that all are linear models. This is discernible by the fact that none of them involves powers or products of the dependent variable y or its derivatives. Linear differential equations are the only ones for which a complete analytical theory exists and for which general analytical solutions can be obtained. Most procedures for solution of non-linear equations consist of 'linearizing' the equation and then using one of the standard techniques for obtaining solutions.

Scheidegger (1970) points out that the above models should account for the lowering of slopes normal to the surface rather than vertical. The geometry involved leads to non-linear equations which have to be solved by different techniques. For example, where $\partial y/\partial t$ is made a function of height, the corresponding equation is

$$\frac{\partial y}{\partial t} = y\left[1 + \left(\frac{\partial y}{\partial x}\right)^2\right]^{1/2} \tag{7.7}$$

which is non-linear because the derivative $\partial y/\partial x$ is squared. In obtaining solutions the differential equation is converted into a difference equation and this is solved on the computer. The results for this model are given in Scheidegger (1970, p. 140).

One may ask what happens at the crest and at the stream channel? This raises the last general point concerning differential equations, that of boundary conditions. The above solutions were general in that (except in one case) the initial conditions and parameters were not specified. Obviously, in interpretation we have to have real, particular values for the solutions. In addition to providing the initial conditions and any constants we have to specify the conditions at the boundaries, i.e. the values which must obtain at two or more values of the independent variable. For example, we could assert that in the above situation a boundary condition is that $y=0$ at $x=0$, i.e. base-level is at the foot of the slope. Another one, very familiar to geomorphologists, is that velocity is zero at the bed of the stream i.e. when y, the distance above the bed is zero. Obviously, the boundary conditions may themselves be time dependent; thus, the value of y at $x=0$, the height of the stream channel, could be lowering through time, for example in a simple linear fashion (Culling 1963). Where boundary conditions obtain they have to be incorporated in the solution, not simply qualified after the general solution is obtained.

Equations of systems

We have already outlined the types of differential equations which exist, partial and ordinary, linear and non-linear, and first, second, third order. Linear systems without feedback may be defined in terms of the order of the differential equation which describes the system's operation with respect to time.

In a zero order system, using the terminology of Grodins (1963), the response or output from a process (forcing function, input) is independent of time, so that a time derivative is absent from the equation describing the system. Thus a zero order system simply multiplies the input by the transfer function (which in this case is simple gain) but does not change the timing between input and output which is one of instantaneous response.

In a first order system, the equation is differential of the first order, so that the response to a sudden increase in the independent variable (forcing function) produces a transient response which gradually damps down to a fixed level which is steady. The situation in steady conditions is the same as for the zero order system, time independent. The period before this is the transient phase. The output no longer follows the input instantaneously, so its value depends on the time at which it is measured. An example in geomorphology is provided by the effect of precipitation (forcing function) on the soil moisture (output) by the transfer function represented by the infiltration rate. Soil moisture in the earlier stages of a uniform continuous rainfall will increase in the transient stage until it gradually reaches a steady level determined by the hydraulic conductivity and infiltration capacity rate of the soil (fig. 7.5).

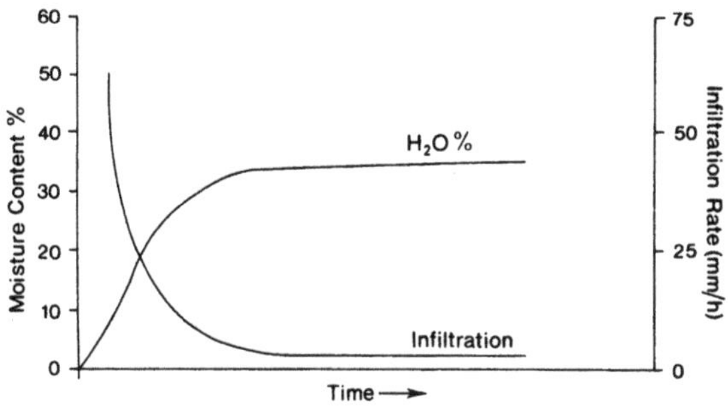

7.5 Hypothetical infiltration response to precipitation.

In a second order linear system, the derivative in time is second order in at least one term. Typically, it should take the form

$$b_1 \frac{\partial^2 x}{\partial t^2} + b_2 \frac{\partial x}{\partial t} + b_3 x = c \tag{7.8}$$

The existence of the second order term, taken together with the first-order term, implies that the response will be damped to a steady state again, but that with particular values of the coefficients b_1, b_2 and b_3 the system may (a) reach equilibrium like a first order system, or (b) reach the steady state level by a series of damped oscillations. An example of this is shown in fig. 7.6 which illustrates the damping of groundwater inflow to a stream caused by passage of a flood wave. Another important case is the passage of a surface temperature wave (positive or negative) into the ground. This takes the general form of a damped sine wave curve with depth. So far little application has been made of second order models in geomorphology, though there are several areas in which they might be expected. One is in the study of sea-level fluctuations, where damping of climatic oscillations and the glacier response seems to have been important. Again, the pattern of isostatic rebound might imply the existence of a double energy storage phenomenon which is characteristic of these second order systems.

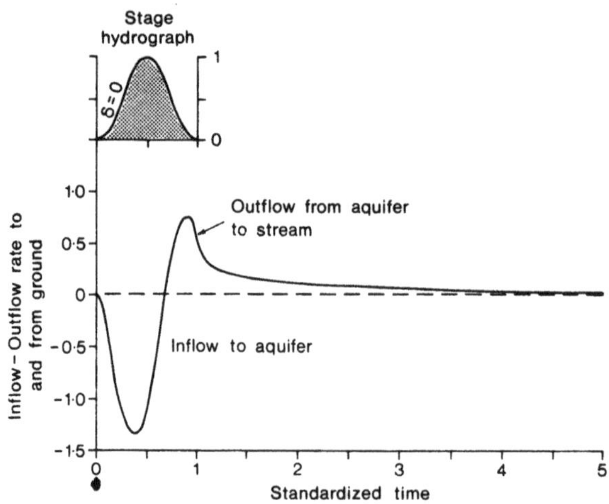

7.6 The damping of groundwater inflow and outflow to a steady state condition caused by the passage of a flood wave in a stream channel (after H. H. Cooper, Jr, and M. I. Rorabaugh 1963).

Equations of continuity and diffusion

One important component of equations describing the behaviour of continuous matter is the requirement that all mass is accounted for. Put at its crudest level, this could be called the 'what-goes-in-must-come-out' equation. Together with an equation of motion, an equation of state, a kinematic condition and the appropriate initial and boundary conditions it provides a complete description of the behaviour. In all geomorphological situations mass is being moved from one position to another, whether soil, water, solid rock, channel debris, or solutes mixing in a stream. The conservation of mass is so absolutely fundamental that it forms the core of most physical models. It is sometimes called the equation of continuity and has two expressions, one representing steady flow of an incompressible fluid and known as the Laplace equation; the other a time-dependent flow, occurring before steady flow is reached and generally known as deterministic diffusion. The last term is to differentiate it from probabilistic diffusion. In the deterministic model we expect to find a partial derivative with respect to t, and indeed this is the case.

If we imagine a simple cell, whose three axes are Δx, Δy, and Δz (fig. 7.7) we can use geometry and some simple symbols to obtain the steady-flow model. The mass flow into the left-hand side of the cube is given by $M_x = \Delta y \Delta z \rho_L \upsilon_L$ where υ is the velocity of flow through that face, $\Delta y \Delta z$ is the area of the face and ρ is the density. A similar expression can be obtained for the right hand face and the difference between them given by:

$$\Delta M_x = \Delta y \Delta z \rho_R \nu_R - \Delta y \Delta z \rho_L \nu_L \qquad (7.9)$$

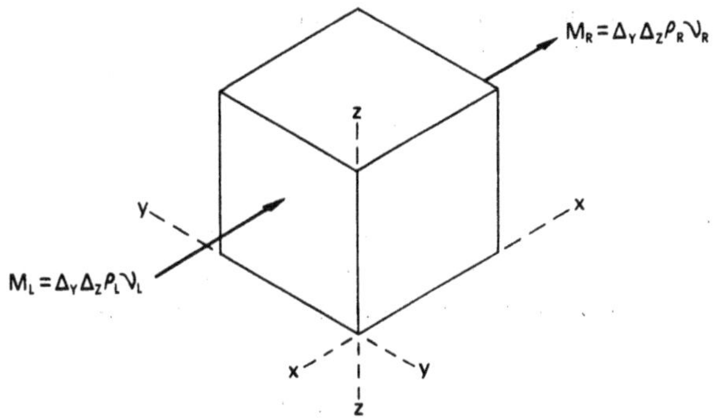

7.7 Notation for development of steady state diffusion model.

which, taking out the common elements and letting Δ again mean 'difference' we have:

$$\Delta M_x = \Delta(\rho v)_x \Delta y \Delta z \quad (7.10)$$

This is also true in the y and z directions and the notation becomes:

$$\Delta M_y = \Delta(\rho v)_y \Delta x \Delta z \quad (7.11)$$

$$\Delta M_z = \Delta(\rho v)_z \Delta x \Delta y \quad (7.12)$$

Now the conservation equation says that:

$$\text{input} - \text{output} = \text{accumulation}$$

In this case accumulation is represented by the change in mass of the fluid element, and hence of average densities ρ_1 and ρ_2 over a short period of time. The equation then is

$$\text{input} - \text{output} = (\Delta M_x + \Delta M_y + \Delta M_z)\Delta t \quad (7.13)$$

that is, the change in flow across the faces in a unit of time, and accumulation $= (\rho_1 - \rho_2)\Delta x \Delta y \Delta z$. Thus letting $\Delta\rho_t = \rho_1 - \rho_2$ we have the unpleasant looking equation:

$$[\Delta(\rho v)_x \Delta y \Delta z + \Delta(\rho v)_y \Delta x \Delta z + \Delta(\rho v)_z \Delta x \Delta y]\,\Delta_t = \Delta\rho_t \Delta x \Delta y \Delta z \quad (7.14)$$

and if we divide through by $\Delta x \Delta y \Delta z \Delta t$ we are left with:

$$\frac{\Delta(\rho v)_x}{\Delta x} + \frac{\Delta(\rho v)_y}{\Delta y} + \frac{\Delta(\rho v)_z}{\Delta z} = \frac{\Delta\rho_t}{\Delta t} \quad (7.15)$$

If the fluid has constant density, then ρ is constant and:

$$\frac{\Delta v_x}{\Delta x} + \frac{\Delta v_y}{\Delta y} + \frac{\Delta v_z}{\Delta z} = 0 \quad (7.16)$$

as the values are considered continuous and very small, this relationship can be expressed by the partial differential equation:

$$\frac{\partial v_x}{\partial x} + \frac{\partial v_y}{\partial y} + \frac{\partial v_z}{\partial z} = 0 \tag{7.17}$$

Notice that there is no derivative with respect to t, as expected. Now velocity is a function of the change in velocity potential (Φ) in any particular direction, e.g. $v_x = \partial\Phi/\partial x$ and we have (substituting in the above equation) the second-order differential linear equation:

$$\frac{\partial^2 \Phi}{\partial x^2} + \frac{\partial^2 \Phi}{\partial y^2} + \frac{\partial^2 \Phi}{\partial z^2} = 0 \tag{7.18}$$

which is the Laplace equation. The expression for the summed second derivative of a variable in three dimensions, i.e. $\partial/\partial x^2 + \partial/\partial y^2 + \partial/z^2$ is given by ∇^2 called the Laplacian operator, so the above equation can be represented by

$$\nabla^2 \Phi = 0 \tag{7.19}$$

Various analytical, graphical and experimental techniques are used for showing this basic equation. For a unit volume, the expression used above, that velocity is proportional to potential

$$v_x = \frac{-K\partial\phi}{\partial x} \tag{7.20}$$

is used for flow in soil, where Φ= hydraulic potential. With steady flow in a homogeneous, isotropic medium the flow can then be described by Darcy's Law, in which K is the coefficient of diffusion.

$$v_x = \frac{-K\partial h}{\partial x} \tag{7.21}$$

These conditions are relatively rarely encountered in natural soils, but the formulation is important because it allows simple models to be built and forms a bridge to the diffusion models, which are also based on the continuity model, input-output = accumulation. Here, the basic assumption made earlier, that flow is steady and time independent, is relaxed. The concentration of mass in the cube is assumed to vary through time, and of course this change of mass represents accumulation or loss. Assuming again matter which is incompressible (density remains constant) then the change in concentration depends on the mass-flow into and out of the cell through its various faces. If J_x is the net mass-flow in the x direction, then the accumulation is the sum of the net mass-flow through all the faces, expressed as:

$$\frac{\partial c}{\partial t} = \frac{\partial J_x}{\partial x} + \frac{\partial J_y}{\partial y} + \frac{\partial J_z}{\partial z} \tag{7.22}$$

where c = concentration accumulation = input — output.

Now $J_x = -K\partial c/\partial x$ where $\partial c/\partial x$ is the concentration gradient and K is a diffusion coefficient in the x direction. Similar expressions can be obtained for the other directions and if we assume that K, the diffusion coefficient, is constant in all the directions, then substituting in the previous equation, the result is:

$$\frac{\partial c}{\partial t} = -K\left(\frac{\partial^2 c}{\partial x^2} + \frac{\partial^2 c}{\partial y^2} + \frac{\partial^2 c}{\partial z^2}\right) \tag{7.23}$$

or

$$\frac{\partial c}{\partial t} = -K\nabla^2 c \tag{7.24}$$

This important, fundamental differential equation is called Fick's second law of diffusion. Together with the condition of continuity from which it is derived, it forms an important core of mathematical geomorphological theory. As the continuity condition has different formulations, so too does the diffusion equation and its solutions, however the basic form remains essentially that described above. Sometimes the equations are simplified rather than made more complex by the fact that they may be taken in only one or two directions; the initial and boundary conditions still have to be specified. Once again it is important to note that while the formation of the problem into differential equations is the primary field of geomorphological interest, solution of the equations, subject to various conditions, is a substantial task.

In the following paragraphs we demonstrate some 'variations on a theme of continuity' which are used in geomorphology to illustrate the formulation of the problem in continuity terms. Later in the chapter some of these are expanded to illustrate other aspects of deterministic mathematical modelling in time.

An interesting early example of the use of the continuity idea in its most elementary form occurs in Fisher's (1866) paper on the recession of a rock cliff associated with build-up of a scree beneath. The mass to be conserved is the rock which changes density and piles up at the foot of the slope. The model was subsequently developed by Lehmann (1933) and Bakker and Le Heux (1952). The basic continuity expression allows for change in the volume as debris is produced by the formula

$$\frac{V_R}{V_D} = (1 - c)$$

where c is a constant and V_R is volume of bedrock and V_D is the volume of debris.

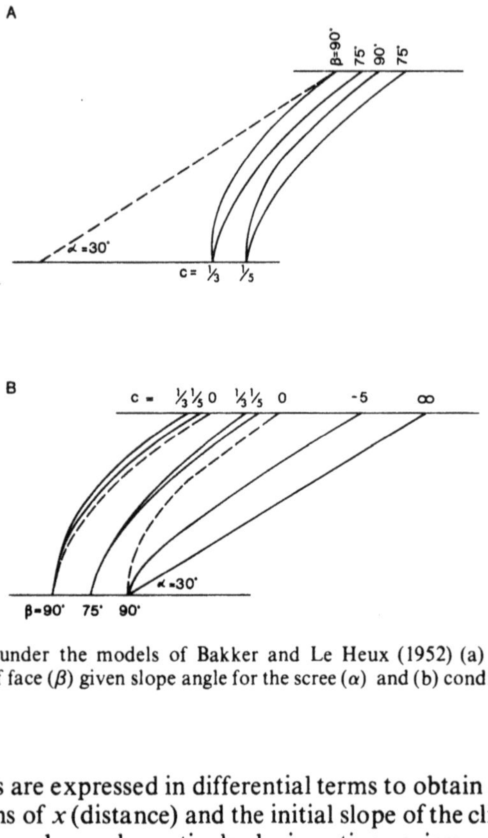

7.8 Cliff recession under the models of Bakker and Le Heux (1952) (a) for various initial conditions of the cliff face (β) given slope angle for the scree (α) and (b) condition of volumetric change (c).

These volumes are expressed in differential terms to obtain an equation for y (height) in terms of x (distance) and the initial slope of the cliff-face, β. The general solution can be made particular by inserting various values of α and c and some solutions are shown in fig. 7.8.

Another continuity formulation for the conservation of ice-mass in an infinitely wide glacier is given by:

$$\frac{\partial q}{\partial x} + \frac{\partial h}{\partial t} = b \tag{7.25}$$

which is the fundamental starting point for study of the motion of a glacier. In this equation the net mass balance (b) is equal to change in the depth of ice flow (h) + the change in ice discharge q; while x is the coordinate direction. An almost identical expression:

$$\frac{\partial y}{\partial t} + \frac{\partial q}{\partial x} = i - f = i_0 \tag{7.26}$$

(Eagleson 1970, p. 332) may be used to express continuity in overland flow. If $i - f = i_0$ where i = point rainfall intensity, f = infiltration rate and

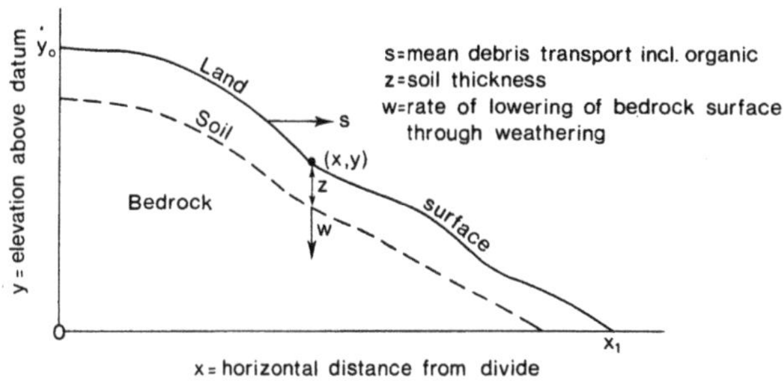

7.9 Terminology for formulation of the continuity equation (after Carson and Kirkby 1972).

i_0 = rainfall excess intensity, then this equated with the change in flow depth in channel (y) + the change in overland flow discharge per unit width of channel (q).

A third example of the formulation of a continuity expression is from Kirkby's work on hillslope processes (see Carson and Kirkby 1972). Kirby's expression, for which the terms are shown in fig. 7.9, is: debris transport in — debris transport out over a unit length of slope profile — increase of soil thickness due to weathering and addition = decrease in elevation of land surface. The difference in debris transport (in and out) is expressed as $\partial s/\partial x$; the change in soil thickness due to weathering is $(1 - \mu) \,.\, W$, where μ is a constant similar to Lehmann's c (Lehmann 1933), and W is the weathering rate; the ground loss in time is given by $-\ \partial y/\partial t$ (being negative to indicate ground loss). The full expression is then given by:

$$\frac{\partial s}{\partial x} - (\mu - 1) \,.\, W = -\frac{\partial y}{\partial t} \tag{7.27}$$

A second continuity equation in Kirkby's work relates to change in soil thickness, thus in differential terms:

$$\frac{\partial z}{\partial t} = \frac{\partial y}{\partial t} + W = \mu \,.\, W - \frac{\partial s}{\partial x} \tag{7.28}$$

where $\partial z/\partial t$ is increase in soil thickness.

The diffusion equations were first developed for work on heat conduction, and it is mainly through soil temperature and glaciology that they make an appearance in geomorphology. On a glacier the surface cold wave is transmitted down into the glacier. The annual 'wave' may be treated in this fashion, and it can be described by the expression:

$$T_t = T_s \sin Wt \text{ (boundary condition, } y = 0) \tag{7.29}$$

where T_t is temperature, t time, W_t the frequency of temperature change at

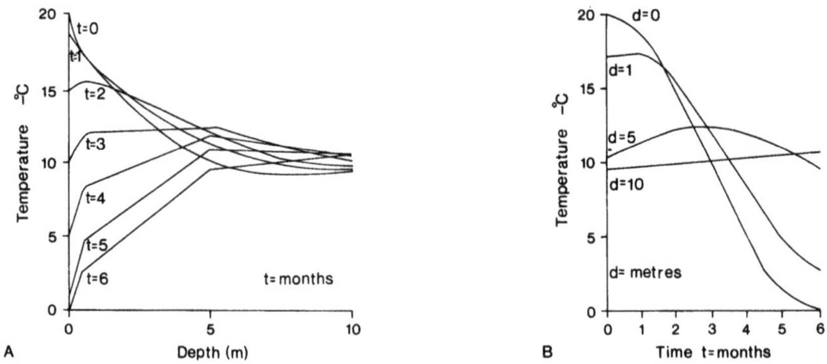

7.10 Decay of thermal effects of a positive increase in temperature at the surface of a glacier as the temperature wave is transmitted down through the ice. (a) The variation in depth for fixed time periods. (b) The variation in temperature for fixed depths. Notice in (b) how the positive increment in temperature at Time = 0 reaches d = 5 after about 3 months. Below d = 10 there is little change with time (from data in Paterson 1969).

the surface, T_s the amplitude of the surface temperature wave and

$$\frac{\partial T}{\partial t} = K \frac{\partial^2 T}{\partial y^2}$$

$\partial T/\partial t = K[\partial^2 T/\partial y^2]$where y is depth in the ice and K the thermal diffusivity coefficient. Thus if a solution can be obtained, the result will give the temperature at depth y and time t, expressed as $T(y,t)$ and given by the equation:

$$T(y, t) = T_s \exp[-y(W/2K)^{1/2}] \sin[Wt - y(W/2K)^{1/2}] \qquad (7.30)$$

when transient behaviour (in the settling-down period) is ignored. The result is shown, using $K = 38\text{m}^2 y^{-1}$, $W/2\pi = 1y^{-1}$ in two forms, in fig. 7.10, and follows Paterson (1969).

Scheidegger (1970) uses the basic analogy of height and temperature to propose a diffusivity equation of landscape decay. The two-dimensional expression for thermal diffusion is given by—

$$\frac{\partial T}{\partial t} = \text{D} \left| \frac{\partial^2 T}{\partial x^2} + \frac{\partial^2 T}{\partial y^2} \right| \qquad (7.31)$$

where T is temperature, x and y spatial coordinates and D a diffusion constant. Substituting h for T (height for temperature) and considering the one-dimensional case the solution of the equation gives

$$h = \frac{1}{(4\pi Dt)^{1/2}} \exp\left(-\frac{x^2}{4Dt}\right) \qquad (7.32)$$

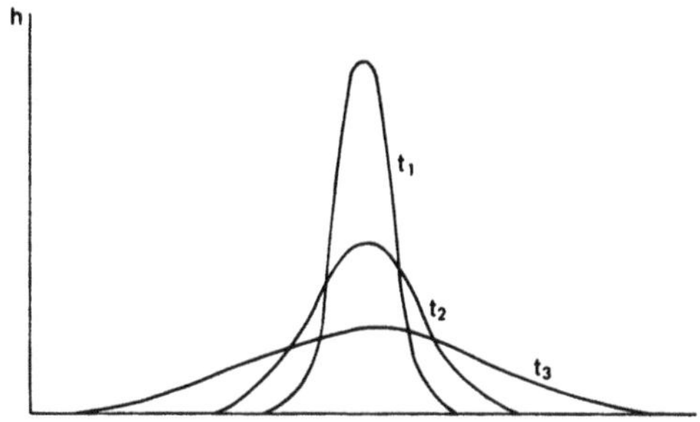

7.11 Decay of a mountain under the assumptions of the diffusion model for three successive time periods (after Scheidegger 1970).

which when plotted for three values of time yields the results shown in fig. 7.11.

As was mentioned earlier, the continuity equations represent only one of the elements required to describe the behaviour of matter. The other essential components in most dynamic time models are the equations of motion. These take on an even wider variety of forms, and will not be discussed at length here. Instead, three interesting and important models for change through time will be used to demonstrate the wider aspects and implications, as well as procedures, for this type of modelling.

Kinematic waves

An important body of theory and technique has been developed since 1955 for dealing with zones of higher concentration in flows. The theory, as developed by Lighthill and Whitham (1955), had been described earlier by Massau (1889) and Seddon (1900) as it applied to flood waves. In essence the problem is this: given movement of some matter in a channel how may the passage of a zone of higher concentration be studied; what are the characteristics of this wave of higher concentration — how fast does it travel and under what conditions is it dissipated?

Consider a few examples to see what 'concentration' means in this context. Cars moving down a single lane on a road are variably spaced; the closer they are together, the greater the concentration per unit length of road. It is common experience when driving on a highway that groups of high concentrations may be passed through and left behind. Every care moves *through* the concentration because it (the concentration) moves down the road at a slower rate than any individual vehicle. The high concentration is called a kinematic wave. Another example is the movement of a flood-wave in the river. Here concentration is expressed in terms of volume of water per unit length. If the channel were of fixed shape, then this concentration will be represented by the height (h) (flow depth in channel). In practice the channel changes shape, but (h) varies more smoothly than cross-sectional area and so is preferred. In yet another example, Langbein and Leopold (1964) described concentrations in dunes and riffle bars in flumes and rivers in terms of the weight per unit distance above the plane representing the base of the moving feature. Finally Nye (1958) described a model in which a kinematic wave is effectively represented by a change in height along the glacier, except at the snout where it is converted into a forward motion.

If one-dimensional flow systems are considered, then kinematic waves occur if there is a functional relationship between flow (g) which is the quantity passing a given point in a unit time, concentration (k) which is quantity per unit distance and position x. This relationship is provided by a continuity equation and an equation of motion, which are combined to describe the path taken by the 'hump' or wave through time and space when they are solved. The basic continuity equation (using h = flow depth in channel) is:

$$\frac{\partial h}{\partial t} + \frac{\partial q}{\partial x} = i_0 \qquad \text{from (7.26)}$$

or

$$\frac{\partial h}{\partial t} = i_0 - \frac{\partial q}{\partial x} \qquad (7.33)$$

where i_0 is the net inflow rate to the surface runoff system or rainfall excess. This states that the change of water depth ($\partial h/\partial t$) with time on a hillslope over a very short section is equal to the rainfall excess, minus the inflow-outflow difference over this short section (p. 146). Here concentration is measured by depth, in a uniformly wide channel with steady flow. Secondly,

$$q = \alpha h^m \qquad (7.34)$$

relating discharge (q), to concentration (h). Where α is a kinematic wave parameter ($L^2 t^{-1}$) and m is a dimensionless kinematic wave parameter according to surface roughness. Horton (1939) set $m = 2$ for most natural surfaces. It is found that, using the method of characteristics, a solution for this pair of equations is given by:

$$c = \frac{dx}{dt} = \alpha m h^{(m-1)} = mV \qquad (7.35)$$

where α, m are kinematic wave parameters regarded in overland flow as constant and V is stream velocity. $c = dx/dt$ is the speed of propagation of the kinematic wave (=flood) downslope and it can be seen (where $m = 2$) that it propagates about twice the speed of the velocity of the flowing water. The corresponding equations for the glacial case are, for continuity:

$$\frac{\partial q}{\partial x} + \frac{\partial h}{\partial t} = b \qquad (7.36)$$

and

$$q = \mu h \qquad (7.37)$$

as defined earlier, where μ is the velocity averaged over the ice thickness h.

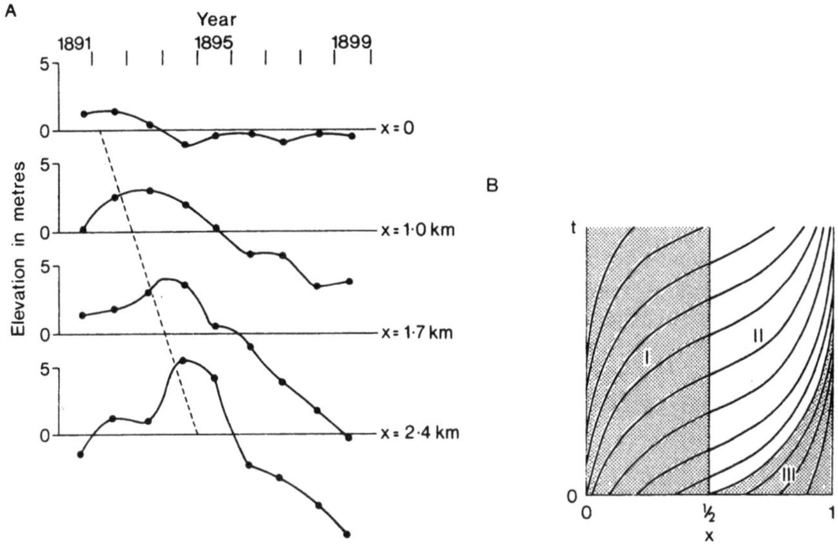

7.12 (a) Passage of a surface wave of ice down the Mer de Glace (after Paterson 1969, from Lliboutry 1958). Each graph shows, for a cross section at distance x, the relative height of the glacier through time. Zero level is arbitrary in each case. (b) Wave paths, in the x, t plane for an idealized glacier (after Nye 1960).

The problem here is to determine how and over what period of time an addition (excessive snow fall period) to the 'normal' steady-rate regime could be transmitted down glacier as a kinematic wave, and eventually reach the snout. The passage of a wave of elevation down the Mer de Glace in France over a period of 9 years was described by Lliboutry (1958) and is shown in fig. 7.12a. With some manipulation and simplification and substitution of values for the speed of the kinematic wave from real glaciers, models describing the down-glacier passage of the wave can be obtained, again by the method of characteristics. The characteristics represent the paths of kinematic waves in the time and space plane and are shown in Nye's (1960) case for an ideal glacier in fig. 7.12b. In this diagram x is distance down glacier, which stretches from 0 to 1, $\frac{1}{2}x$ being on the line between the accumulation and ablation zones. The lines would therefore 'fix' the migration of a kinematic wave at any point in time (y axis) or space (x axis). The characteristics, together with some assumptions about strain rates in glaciers, can be used to predict the expected response times to a set of impulsed inputs. This analysis is however further complicated by the fact that the kinematic waves themselves undergo diffusion and so their forms diminish. This in turn affects the response rate and in fact greatly lengthens it, by slowing down the rate of adjustment and increasing the amount of adjustment needed. Finally Nye (1965) considers the response time through the transfer function approach outlined earlier.

Hillslope process-response model

Recall (p. 147) that Kirkby set up the continuity equations

$$\frac{\partial s}{\partial x} - (\mu - 1)W = -\frac{\partial y}{\partial t} \qquad \text{from (7.27)}$$

and

$$\frac{\partial z}{\partial t} = \frac{\partial y}{\partial t} + W = \mu W - \frac{\partial s}{\partial x} \qquad \text{from (7.28)}$$

where s = debris transport, z = soil thickness, y = elevation above datum and W reduction of elevation of bedrock due to weathering. If removal takes place under weathering limited conditions, $\partial z/(c)t$ is equal to zero and $\partial y/\partial t = -W$; i.e. hillslope lowering depends on rate of weathering. Alternatively, if transport is limited ∂s is at capacity rate $\partial z/\partial t$. In the latter case, by eliminating W the continuity equation becomes

$$\frac{\partial c}{\partial x} = -\mu\frac{\partial y}{\partial t} + (\mu - 1)\frac{\partial z}{\partial t} \qquad (7.38)$$

Assuming that $\mu = 1$, $\partial y/\partial t = -\partial c/\partial x$ and the rate of transport is directly related to spatial variations (in one dimension) of transporting capacity. If it is further assumed that there is no threshold slope angle, that unimpeded basal removal occurs and that the initial slope is straight, solutions to the differential equation can be found by the separation of variables. Debris transport rate (c) is related to a function of distance from the divide and some power of local slope; i.e.

$$c = f(x)\left(-\frac{\partial y}{\partial x}\right)^{n} \qquad (7.39)$$

This can be differentiated and substituted into the continuity equation and a solution found (Carson and Kirkby 1972, Appendix B). This solution (which is a purely mathematical solution to the general equation set up following the method of separation of variables) is subject to some approximations which yield, when $f(x) = x^{m}$

$$y = y_0\left[1 - \left(\frac{x}{x_1}\right)^{([1-m]/n+1)}\right] \qquad (7.40)$$

Where y is elevation, y_0 elevation of divide, x is horizontal distance from divide, x, is total horizontal distance and m and n are exponents. This solution Kirkby calls a 'characteristic form' because it depends on the process and not on the initial form and since with the passage of time it changes hardly at all. Substituting different values of the parameters m and n yields particular solutions to this general form. Notice that when $m = 1$, $y/y_0 = 1 - (x/x_1)$ which is a straight line. This occurs of course when the capacity for transportation is a simple linear function of distance from the crest, regardless of n. The forms have been standardized by taking y/y_0 and x/x_0 so that the vertical axis represents the ratio of local height to height at the divide, and the horizontal axis the ratio of local distance from divide to greatest distance from divide. By substituting appropriate values of m and n

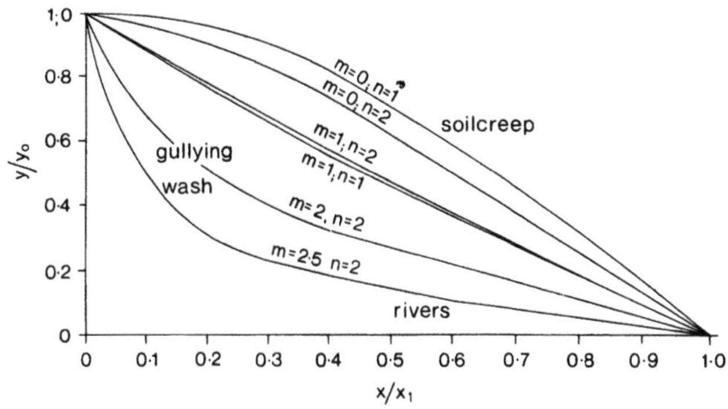

7.13 Solutions to the Kirkby (1971) model according to different values of the parameters m and n. Distances on x and y have been standardized.

(obtained from emperical data) and substituting these in the equation (again with a fixed divide and outlet), the characteristic forms for various slope processes are obtained. This diagram is reproduced in fig. 7.13.

Stability analysis

The last example discusses some recent, experimental work by Smith and Bretherton (1972). The problem is to obtain solutions to equations describing the effects of perturbations on (1) a surface which is initially fairly smooth,

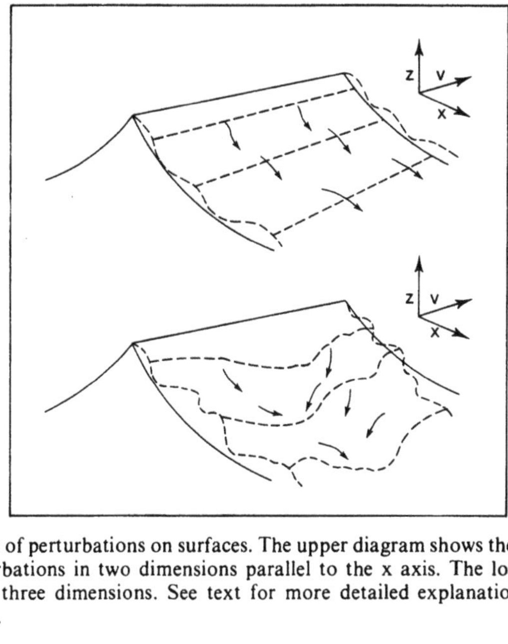

7.14 The effect of perturbations on surfaces. The upper diagram shows the effect on a smooth slope with perturbations in two dimensions parallel to the x axis. The lower diagram shows perturbations in three dimensions. See text for more detailed explanation (after Smith and Bretherton 1972).

and (2) a v-shaped channel system. The procedure is to set up a model using basic continuity laws and equations of motion. This model, represented as usual by partial differentiation equations, is then subject to a perturbation. In other words, in the drainage basin, which is everywhere in steady state, the surface is instantaneously modified by a small amount. If this small perturbation gradually disappears when the model is re-started, then the basin is said to be stable. If, on the other hand, after re-starting the model the perturbation (i.e. depression or knickpoint) begins to grow, then the basin is unstable. This technique unfortunately applies only to small perturbations, so that if the system is unstable, the assumptions of the technique prohibit us from following through the evolution of the system for any length of time.

In their first experiment, Smith and Bretherton consider perturbations in two dimensions only, parallel to the x axis (fig. 7.14). In this case, perturbations are removed because an increase in slope causes an increase in sediment transport, whereas a decrease in slope results in decrease in transport. One implication is that knickpoints will always be removed by migration 'up-gradient', an observation which is supported by flume work. The second experiment relates to perturbations in three dimensions. Given their transport law, it is concluded that with a constant form surface which is elsewhere concave, there can be no stable channel system on the surface; if it is straight or convex there is no instability and channels cannot develop from small perturbations. With a landscape combining the two elements, one part (the convex) would inhibit channel development (negative feedback) whereas in another area it would be unchecked (positive feedback).

This idea of stability and instability through space is quite important, and conforms with the idea of a transient spatial behaviour comparable to transients in time. It is in the linking of spatial and temporal components that the full benefits of deterministic mathematical modelling will be reaped. At the moment, as this brief review has shown, the difficulties of obtaining analytical solutions for models which are cast in several dimensions and especially those which are non-linear are very considerable and, as yet, the amount of three-dimensional analytical modelling is very modest. Some of these problems may be overcome by simulation (see p. 158); but at present, despite these problems, analytical modelling is the most powerful tool of theoretical geomorphology.

8 Stochastic models of form evolution

Stochastic models of change are models in which there is some specified amount of uncertainty about the results taken on by particular variables which represent the output of the model. For example, suppose we wish to model the processes operating along the main channel in a semi-arid gully system and that this model requires flow from the tributary channels. The model-builder might incorporate a degree of randomness in the nature, timing and magnitude of discharge events in the tributary systems in relation to supposed precipitation-event frequencies. In this way he builds uncertainty into the model.

In the preceding chapter we sought perfect prediction of geomorphological phenomena using acceptable theories formulated in terms of deterministic laws. Why then should we actually incorporate uncertainty? There appear to be two arguments for doing so. First, it is argued that the natural world, being the result of a long period of changing processes, on various lithologies and under different environmental conditions, is infinitely variable. As Melton (1958) puts it, 'to argue that this variability could ever be completely explained is absurd'. In other words, the inherent complexity means that it is unlikely that 'pure' models can ever be matched in the real world. Davis, the author of the cyclic model of landform evolution, similarly recognized that the length of time required and the complexity of the real world might prevent the development of a perfect peneplain. As a result of this complexity it is argued that only 'average' or normative statements, which are of a statistical character, may be made about processes and the outcome of processes. It is readily admitted that 'Each process is deterministic; however the rates and periods differ and the results may be indistinguishable from random' (Scheidegger and Langbein 1966). Recognition of this basic character of natural processes has afforded an alternative line of attack to the procedure of deterministic model-building. Instead of seeking a very complex model structure, the stochastic model-builder starts with extremely simple assumptions and gradually imposes constraints on the freedom of action of his random processes.

The second reason for utilizing stochastic processes is that inputs to many models really are, as far as our world is concerned, randomly variable. Certain input variables have relatively short 'memories' such as the impact velocity of a raindrop on a point. Any particular raindrop is independent of the previous one in this respect. Others have intermediate persistence, for example the amount of rain in 24 hours will perhaps be related to that of the previous and succeeding days in an area of temperate cyclonic disturbances and so on. These phenomena can be described in terms of their trend, seasonal and random components as outlined earlier (chapter 4) and such effects can easily be incorporated to add realism to the models.

Where a random component is involved it often creates problems in the verification and especially the validation of the model. The principle of these is that any specific result, by virtue of the nature of the process, may represent the 'tail' of a distribution of possible sample results and hence be a relatively extreme case. For this reason an ensemble of results has to be obtained before normative statements may be made. If the average result does not match up with the contemporary landscape, we must then ask whether the model is wrong or whether we are comparing it with an 'extreme' case in the contemporary landscape. In deterministic modelling, because uncertainty in each particular component of the model is not allowed, the premises on which they are based are either correct or incorrect. In stochastic modelling there are constraints on the degree of uncertainty, but emphasis is on the 'aggregate' success of the model which does not stand or fall by its component parts.

Types of stochastic modelling

There are five basic types of stochastic model. The first and crudest level of employment of chance components into a model is used where the values of a single variable in an otherwise deterministic model are supplied by a generating process controlled by a statistical distribution. This is called the Monte Carlo technique. This technique may be used to set up the starting values for a model which then proceeds to evolve in a deterministic fashion. For example, we may wish to assign values of height to a hypothetical surface to create roughness which is subsequently to be used in a drainage development model. Once these height values have been selected there may be no further demands on the generating technique and no further randomness is incorporated in the model.

Secondly, values may be required for the variables at regular intervals during the operation of a model and after these have been supplied adjustments take place within the model according to a fixed set of rules, the model 'evolves' and then new values are supplied. This is a common form of digital probabilistic simulation, usually based on a computer-operated model in discrete time intervals.

On a third level, truly stochastic processes are involved, in which mathematical techniques and assumptions are incorporated into the model and provide the uncertainty from within, rather than simply providing it from an external source. This approach is normally employed when the values supplied for any one variable are not independent of each other and includes models such as the simple independent event random walk model and the Markov Chain structure.

On the fourth level are more complex stochastic processes involving continuous time and state space. So far in geomorphology these have been used only in a theoretical manner because of the difficulties of obtaining suitable parameter values for the models and obtaining solutions to the equations. Examples are the complex three-dimensional stochastic diffusion models of the type employed by Culling (1963) and Scheidegger (1970).

On the fifth and most abstract level, are the entropy maximizing models which have recently been borrowed from statistical mechanics. These models, still essentially probabilistic in nature, rely on very different and rather more fundamental propositions than the earlier types described, though in fact they were among the first stochastic models used to describe geomorphological phenomena; notably by Von Schelling (1951) for meanders.

Monte Carlo techniques

Imagine a model which is to test theories relating to the mechanics of thaw and expansion of thermokarst lakes. One input component of such a model could be wind speed and direction. Assume that we have 'runs' of wind from one direction, i.e. it flows constantly for some period from one direction and then constantly from another. Suppose the remaining mechanics of overturning, wave development, thermal change and lake growth are determined. The problem then is to provide direction, velocity and period to the model in a fashion compatible with nature. We will illustrate only the generation of wind speeds.

Column 1 gives wind speed in the Beaufort Scale. Column 2 gives the probability of the wind speed equalling that of Column 1. In the third column

x	$P(W=x)$	$P(W \leqslant x)$	Random numbers
0 --1	0.210	0.210	000 — 209
1+ --2	0.350	0.560	210 — 559
2+ --3	0.200	0.760	560 — 759
3+ --4	0.100	0.860	760 — 859
4+ --5	0.040	0.900	860 — 899
5+ --6	0.034	0.934	900 — 933
6+ --7	0.036	0.970	934 — 969
7+ --8	0.030	1.000	970 — 999
	1.000		

is the cumulative distribution, giving the probability of obtaining a speed equal to or less than that in the first column; wind speeds here have never exceeded 8 during the period of record. These are the prior probabilities. If we now allocate blocks of numbers between 0 and 999 as in column 4, we may use a random number generator to pick numbers at random between and including 0 and 999. Using these, we can draw a series of wind speeds by referring to the appropriate sub-set in column 3. We would use similar devices to obtain direction and period. This technique is suitable for variables which can be allocated to discrete classes. In this case we might be forced to block out our period variable into 30-minute intervals, which could be inconvenient. Continuous process generators are therefore used to overcome this problem by using the equation for the cumulative probability function.

Digital stochastic simulation

The Monte Carlo technique provides the core of most simple digital simulation models and most simulation models of systems are referred to as 'Monte Carlo models' where they involve probabilistic operations of the type described earlier.

The basic reasons for adopting simulation techniques in geomorphology are:

(i) Many systems or sub-systems of geomorphological interest cannot directly be observed without interference. For example, in soil creep, channel processes and wave phenomena the introduction of observational instruments interferes with the processes in operation.
(ii) The rate of operation of geomorphological processes is extremely slow. It is argued that if we develop a satisfactory model (probabilistic or deterministic) we may overcome this difficulty in part.
(iii) In constructing the simulation model, the builder is obliged to direct attention to the areas of weakness, i.e. it highlights inadequacies.
(iv) It allows scale reductions in time as well as space.
(v) It facilitates the elaboration of specific models relating to geomorphological development and process operation.

It is this last use which is by far the most important aspect of simulation work and which most adequately demonstrates its potential. Here simulation is used as an adjunct to research in a way that is similar to the use of multivariate analysis as a search procedure. In the latter the objective is to minimize the unexplained variance. In simulation the most successful models are said to be those that come closest to 'reality'. Unfortunately, the problem of verification, mentioned earlier, again looms large and for three reasons. First, we have to have a consensus as to what constitutes reality, and this implies a much higher level of information than at present. Secondly, a decision has to be made regarding the criteria on which the closeness of fit between the real world and the simulated model is to be evaluated; and thirdly, any given landform in the real world may result from a range of possible 'causes'; in effect this is the equifinality problem mentioned in the introductory chapter.

In using simulation for short-term forecasting, the problems are less severe. Here, though, the objectives are different. We can forecast water quality values or discharge values on the basis of simulation of the historical record with only a relatively modest understanding of the mechanisms and, provided that the extrapolations fit the ensuing record, they are effectively validated as forecasting models. This is, of course, the common base of many rules of thumb. In chapter 4, where decomposition of serial records was discussed, the disaggregation leads towards a basic simulation model for the extrapolation of future values.

With the exception of this use of simulation, there must still be some considerable reservations as to the utility of the procedure. As a research and model-building procedure it is a 'tool' to lay alongside regression, Fourier transformation, entropy maximization and so on. Until the difficulties of validation are more fully investigated and the fascination with technology rather than with geomorphology is overcome, an attitude of cautious optimism only can prevail. Other forms and uses of simulation, for example in the solution of equations, are already tried and proved. This is not yet the case in geomorphology.

There have been several hundred published and mimeographed papers on simulation using basic Monte Carlo techniques in topics such as drainage pattern development, hillslope evolution, the migration of coastal dunes, the development of scree slopes, the growth of limestone cave systems and the growth of coastal spits. In fact there are very few areas in which such models have not been applied. Two models are considered here because they illustrate quite well the general procedures and because both are concerned with change through time (as opposed, for example, to random allocation procedures).

The first is Howard's (1971) simulation of stream capture (fig. 8.1a). Simulation is employed because this slow process is difficult or impossible to observe on maps or in the field. The simulation is probabilistic because (i) the scale is such that randomness arising from variations in the elevation of interfluves in the field cannot be shown — so it has to be built in, and (ii) in the model spacing of the streams has to be regular; in nature it is more random — to add 'realism' randomness has to be built in. Having made the decision to proceed along these lines, the following steps are adopted which are common to most digital models:

1. set up initial conditions;
2. define operating rules;
3. run model;
4. calculate, interpret;
5. return to 2 and continue until results of 4 are satisfactory.

The initial conditions are set up on standard routines in which a 40×40 matrix is filled with stream channels, one segment to each cell. A simple case of this is to select a point at random in a cell and allow a stream to 'grow' one link at a time, with equal probability of moving in any of four directions, until it reaches another stream or the margin of the matrix. Some streams thus simulated on various mathematical surfaces are shown in fig. 8.1b from Thornes (1971).

Besides setting up the pattern of streams, elevations along each stream have to be set up initially. Since these change, they have to be re-valued

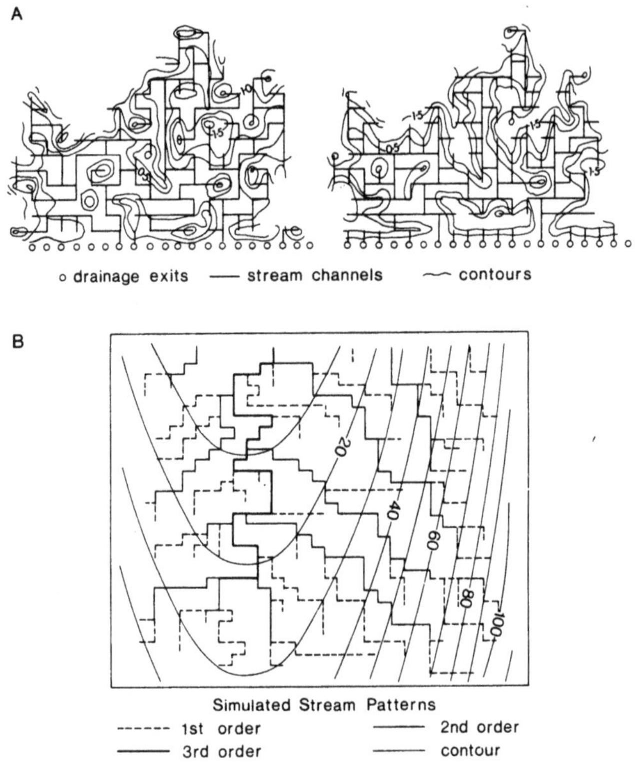

8.1 Examples of simulated stream patterns using Monte Carlo techniques. (a) Two successive stream patterns developed by Howard (1971) using a simulation of stream capture. (b) Stream pattern developed by simulation on a simple topographic surface (Thornes 1972).

during the running of the model. Slopes are made a negative power function of area in the form $S = KA^{-z}$ where S is slope, A is area and z an empirically determined value.

The basic proposition is that the probability of capture increases when the ratio of the slope between two streams to the slope of the stream to be captured increases.

The operation is as follows: (i) a point in the interior of the network is selected at random; (ii) its capture by one of the surrounding streams occurs, or fails to occur according to the rule described in the previous paragraph. To do this the four adjacent cells are examined and the captive stream is taken by the cell with the highest ratio, provided that the ratio exceeds a number generated at random. Various minor rules are inserted to cover all possibilities; for example, if two 'captor' cells have the same probability of capturing the stream under observation then one is chosen at random. Once a capture event has taken place, the network is adjusted, elevations re-computed and the whole process repeated. In fact, to avoid using excessive computer time this operation was done less frequently than initially required by the model.

The validation of the results is expressed in terms of the degree of conformity with the Hortonian characteristics of natural stream channels, and in a qualitative discussion of how various values of z and the initial network configuration affect the results. The author points out that many of the dimensionless properties of natural stream networks are better reproduced by a simulation involving capture than by earlier ones not involving capture, 'suggesting that capture processes may be important in stream network development, especially in the early stages'. The author then gives three reasons for being cautious about this conclusion: (i) the degree of improvement resulting from the capture model is relatively small and might be attributable to chance; (ii) there might be good reasons why natural streams deviate from topological randomness which have nothing to do with capture; and (iii) given that the improvement is very small, it might equally well be related to some other aspect of the simulation, such as the equal spacing imposed by the methods employed.

This paper has been presented at some length because it is a lucid, frank exposition of the techniques and problems involved in digital simulation. Some of these are:

1. simplification of the system to meet computing requirements;
2. a discrete, iterative procedure (i.e. constant repetition of a basic operation) which is jerky and abnormal;
3. complication of the operation by 'minor rules' to cover all computing eventualities;
4. adjusting scale questions and operating rules to meet particular computing requirements;
5. adopting a probabilistic technique when deterministic modelling is impossible;
6. validating the model on a relatively crude set of criteria relative to the question being asked. Does capture take place, and if so, are we any wiser as to the mechanism?

The second case is King and McCullagh's (1971) simulation of Hurst Castle Spit in southern England. Here the authors sought to simulate a specific feature using techniques very similar to those described above. The authors followed a model set up by Lewis (1931) in which wave direction, wave type, wave refraction and depth offshore are all important variables and these have to be incorporated into the simulation. In essence a series of operations fills in a set of cells in the matrix to produce a shape defined by the occupied cells and 'validation' of the model consists essentially of comparison of the shapes. By experimentation a wide variety of conditions could be produced and observations made on the changes in the simulated Hurst Castle Spit which resulted. Whilst the earlier example yielded a large number of simulated drainage patterns whose overall characteristics were examined and then compared with nature, the second example illustrates the use of simulation in a feedback sense. One supposes that after trials which produce an inadequate shape (a poor geometrical fit), the rules as well as the frequencies depths and directions and types of waves also have to be changed and if necessary changed again until a better shape fit is obtained. Eventually the effect of changes in parameters in the model chosen are used to simulate the effects of real world changes.

Stochastic processes

The simulation models just described involve processes whereby the systems involved develop in time and space in accordance with probabilistic laws. Such processes are called stochastic processes. The processes were formalized in terms of computer operations which basically selected random numbers by electronic means. Stochastic processes can be thought of mechanically: the tossing of an unbiased coin, or the drawing of coloured balls from a bag. Finally, they may be formulated mathematically, i.e. we may describe a situation in the real world by a stochastic process which is formulated mathematically.

Consider as an example the random walk, the simplest stochastic process. Mechanically it might be thought of as a thread of water which deviates to left or right of its course by a fixed amount in unit intervals of distance; the probability of a shift to left or right being equal to 0.5 (see fig. 8.2a). In simulation terms, the model could be designed so that it shifts to the left or the right according to a number drawn at random using the Monte Carlo technique. Running this model several thousand times could yield a large number of observations, all of which can be mapped. We could ask the following questions: what is the average number of steps before the thread of water impinges on the left-hand or right-hand bank, for the first time? Or, if we had two streams flowing side by side, what is the likelihood that one stream would 'capture' another by deviating so far that it impinges on the other's path? Every 'run' of the computer model is called a realization of the process. In this process the deviation taken by the stream at each 'turning' point is completely independent of the jump taken at the previous step. Such a process, in which the values straight-on, left or right are the only ones possible, and in which each choice of direction at each step is completely independent of the last, is called a simple random walk. It is not difficult to see how the rudimentary drainage patterns developed for fig. 8.2a are in fact cases of simple random walks with certain simple constraints applied.

Specific stochastic processes of this type can also be formulated mathematically. If X_n is the position after step n of the growing 'thread' of water, then the process is described by the expression

$$X_n = X_{n-1} + Z_n$$

Where Z_n is the jump at the nth step, and every Z is independent of any other step; X_{n-1} is, of course, the position just before the jump is made. For example, suppose when the thread is at position 3 after 6 steps, in the next

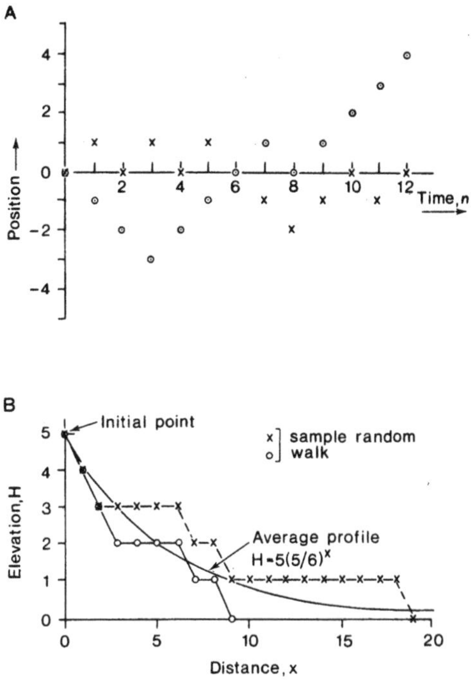

8.2 Examples of random walks. (a) Simple random walks to demonstrate the pattern produced by a stream which is allowed to deviate to left or right by a fixed amount in unit distances where the probability of a shift to left or right is equal at 0.5. (b) The random-walk model of Leopold and Langbein (1962) to develop a series of river long profiles.

step it could move to position 2 or position 4. The move is assumed to be independent of the previous move. It is easy to obtain mathematically, for example, the probability that of 16 steps, 8 or 9 or any number up to 16 are left-hand steps; similarly we could calculate the probability that the 'random walk river' would reach one or other of the confining banks. The possibility of formulating the problem in a purely mathematical way instead of generating thousands of 'streams' and then inspecting distributions of particular values (e.g. number of steps to reach a boundary) leads to enormous economy of computing effort. Unfortunately not all problems can be couched in terms of formal stochastic processes and the number of mathematically simple stochastic processes is small!

Two further characteristics of the simple random walk process need to be noted at this point. The process is discrete in time and in state-space. In this instance the states are represented by the possible positions of the thread. Instead of saying the thread is in position i, we say state i is occupied. The states here are discrete and have sharp boundaries. Discrete-time means that the changes which are occurring in the system occur at fixed time intervals. In this sense it is very close to the simulation models of capture and spit growth.

Random walks of this type have been used in geomorphology both for digital simulation and for deductive model-building, though the latter have been concerned mainly with continuous time and state-space. Leopold and Langbein (1962) used the random walk model to develop a series of river long profiles. They developed the thesis that, without a constraint on river length, the most probable long profile would have an exponential form, i.e. would curve away at first steeply and then more gradually approaching the x-axis at infinity. The model envisages a stream developing in which the system can either (a) remain at the same level, i.e. occupy the same state, or (b) move into a new, lower state. The system is set up such that the probability of a downward step decreases as height above base level decreases. Fig. 8.2b, taken from their paper, shows two realizations of their model and the 'average profile' obtained from 100 runs. Under the specified conditions the most probable profile was negative exponential. The simple random walk is entirely 'memory-free'.

In most geomorphological situations it seems reasonable to expect the 'memory' to be longer, that is to say, to expect that at least successive events will be partially dependent. Systems in which the outcome depends partially on the previous outcome (and not on any outcome prior to that one) are said to exhibit the Markov property. Each result in a succession of experiments is dependent only on the state occupied previously. To illustrate this proposition, imagine a frog jumping about on a set of large lily leaves (fig. 8.3), each lily leaf is a 'state' and when the frog is occupying one leaf, he is said to be occupying that state. The problem could be looked at in two ways. On the one hand, we could watch the frog for a long time and see, each time he jumps, where he jumps to and from. Over a long period of time we could 'tally' the number of jumps between various leaves. For example, he might

8.3 Diagram of a 'probablistic' frog, to illustrate the idea of the Markov property in which each result (position in lilypond), in a succession of experiments (jumps), is dependent on the state (leaf) previously occupied.

jump from leaf A to the various other leaves as follows, in a thousand jumps:

from	to B	to C	to D	to E
A	[200	400	300	100]

Correspondingly, if we see the frog on A then we might assert that he has probabilities of jumping on to B, C, D and E of 0.2, 0.4, 0.3 and 0.1 respectively. In this way, considering him in turn to be on every leaf, we can obtain a matrix of probabilities of jumping from each leaf to every other leaf. This matrix is of fundamental importance in Markov Chain processes and is called the transition matrix, since its entries represent the probabilities of transition from one state to another. If the matrix is multiplied by itself (powered), the probabilities which result are those of going from one state to another in two steps. After repeated powering a situation is reached in which all the rows of the matrix are the same. In other words, the starting point becomes irrelevant after a large number of transitions and repeated realizations of the process would ultimately yield the same distribution of state occupance. This basic chain structure is the root of many quite complex processes. For example, certain states may be absorptive (once they are 'entered' they cannot be left and the realization terminates); others may be made reflexive, if the process enters a particular state, then it returns to the state previously occupied. The reflexive state in the random walk model described earlier occurred when the thread of water 'hit' the valley side wall.

The chain model has not yet been used extensively for modelling in geomorphology, though it seems to have considerable potential. Imagine a set of daily observations on a beach in which beach form is the result of well-defined conditions. These could be swell, onshore wind (state 1); swell, offshore wind (state 2); local sea, onshore wind (state 3); local sea, offshore wind (state 4). Observing these conditions, a transition matrix for a particular coast can be manipulated to yield answers to the questions:

1. In the long term what is the most probable distribution of these four wave conditions?
2. Assuming we find the beach in any particular condition on a day of observation, what is the probability of finding it (i) in the same condition after 1, 2 . . . n days, (ii) in any other state after a specified number of days?
3. If we chose any day, what is the average length of observation (in days) required to observe all or any of the states? What are the variances associated with these means?

It will be apparent that these questions have important implications for experimental design as well as questions of scientific and practical interest. To a large extent the utility of the answers will depend on (i) the value and length of the record on which the transition probabilities are based, (ii) whether or not the observed series is in fact truly Markovian (does the Markov property hold), and (iii) whether or not the series is truly stationary. If the system is stationary this means that the transition probabilities are invariant with time. In this specific model we should want to develop seasonal matrices, representing say winter and summer conditions, since it is obvious that the transition probabilities *will* vary according to season. Soil moisture and its spatial distribution (with stream-basin matrices or matrices representing drainage extension under various storm conditions), are typical of the many discrete time-space observations in which there is relatively low-order persistence. The Markov chain process has been extensively used in geological research in recent years and the reader is especially referred to Harbaugh and Bonham-Carter (1970) for an extended discussion.

The Markov property is also useful in deductive modelling, and indeed it is possible to cast the transition probabilities in a purely deductive framework. In the case of the jumping frog we could, for example, assert that the transition probabilities should be based on the relative distances from each to all the other lily leaves. Consider, for example, the matrix set up by Scheidegger (1966) to describe the manner in which streams might join. If a river segment is of order i it can only join either another stream segment of order i or a stream segment of higher order j, which is greater than or equal to $i+1$. Let the first case have a probability p and the second a probability q. Since one or the other must occur, then $p+q=1$. Now if p and q are not dependent on the order i itself, then the process is Markovian. The matrix appears as follows, with states representing orders

		$j=$ no. of streams of order $n+1$		
		0	1	2
$i=$ no. of streams of order n	0	1	0	0
	1	1	0	0
	2	q	p	0

This is argued as follows, the matrix entry represents the probability of obtaining j streams of order $n+1$, if there were i streams of order n. Now if there were no streams of order n, it is certain that there will be no streams of order $n+1$. Consequently, the transition can only be from zero to zero. If there is one stream of order n there cannot *be* any streams of order $n+1$, so in position (1,0) in the matrix the entry is unity. If there are two streams of order n and they do *not* join, there will be no streams of order $n+1$, and the probability of not joining is q. If there are two streams of order n and one of $n=1$, they must join, the probability of which is p. Finally, with two streams of order n, the probability of obtaining two streams of order $n=1$ is zero. Significantly, the author found this enumeration clumsy and proceeded to use alternative formulations which were more amenable to analytical procedure, but the deductive argument for setting up the transiation matrix is clear.

The final example of a discrete formal stochastic process comes from the theory of queues. Here stochastic processes are applied to solve questions of storage and congestion and they have been widely applied in fields such as telephone engineering, traffic studies and mass-production lines. Intuitively they appear to have considerable potential in geomorphology in situations where we are dealing with storages of discrete items such as rock and soil particles.

Normally the queuing models involve arrivals of the elements, their waiting in a queue followed by servicing and departure. Queuing theory seeks to ask such questions as:

1. If the mechanism for removal is continuously available, what percentage of the time will it be in use?
2. What is the average number in the store or queue?
3. What is the probability of obtaining any particular number in the queue?
4. What is the average time spent in the store by the particle?

Normally one has little or no prior information on the arrival time and the time taken to 'service' an element (i.e. dispatch it from the queue) and these have to be assumed in terms of statistical distributions whose parameters are estimated from the scarce data available. The Poisson process is frequently used for arrivals because it has the important properties that for small time intervals the probability of more than one arrival is very small, and that the inter-arrival time distribution is exponential. The servicing time is described by a negative exponential distribution in the simplest models. The interaction of these two distributions, which are assumed not to change (i.e. are stationary) yields the information described above, if values are provided for the various parameters. In complex models the Monte Carlo technique outlined earlier is used to simulate the queue digitally by providing 'arrivals' and 'departures' and keeping a tally of the size of the queue or store.

Of special interest in geomorphology, however, are bulk queues in which arrivals and services can occur in groups. Thus, when the service is available, if it can service more than exist in the queue, the whole queue will disappear. Removal of gravel stored in a river bed springs immediately to mind. The busy hour phenomenon, in which instead of arrival probabilities being stable they vary with time of day (or year in a seasonally-biased process), is also of considerable geomorphological interest. Finally, queue networks have been studied which bear a very close resemblance to certain geomorphological situations; for example, sediment discharge in a river network. These have the advantage of built-in feedback whereby congestion in the main network causes build-up in the tributary channels. Most studies of this type have concentrated on obtaining steady state solutions which describe the average state of affairs 'in the long term'. In this context Thornes (1971) set up a deductive model of scree slope development under various environmental conditions. Obtaining parameters for the model and efforts to validate it prove singularly difficult, however, because of the paucity of data. Implicit in the notion of steady-state is the idea that the processes have been in operation without significant changes in rate. In terms of most geomorphological processes, which act very slowly, there is only the relatively short period of climatic stability of about four hundred years in which we can assume this condition.

A more useful proposition for a slope transport model based on queuing theory would be to assume that the bottom of the slope (the point of evacuation) comprises a large number of 'outlets' or service points, each of which operates at random, instead of service in order of arrival. This might allow for a more chaotic queue and liken slope particles less to the orderly supermarket queue and more to a particular crowd of various sized football supporters heading for a series of exit gates! Such a model has waiting-time distributions which are much more realistic for natural phenomena where some particles may be moved after a very short stay while others wait a very long time.

Process continuous in time and/or space

The models in which time and state change at sharply defined regular intervals are easier to understand and mathematically less complicated. Furthermore, they lend themselves easily to computer modelling which requires the discrete steps to be programmed. In nature, of course, events are rarely discrete; wind does not blow continuously for blocks of thirty minutes, or even for one minute; streams do not deviate only by one or any multiple of one unit to the left or right; glaciers do not receive 'standard snow fall additions'. There are two approaches to this problem.

First, we can divide the readings into finer classes. For example, for wind direction we may have 360 states which will be quite fine enough to cope with any empirical data! The problem here is that the computation time, effort and cost rises very steeply, especially where matrices are involved. Even so, this expedient often is used as a solution.

Secondly, we may adopt models which are based on continuous processes operating in a fashion unrestricted by artificial time and state barriers.

In our previous examples, if our lengthening wandering thread of water were still observed at unit time intervals, but could move by any amount to left or right (measured perhaps in radians), then the model would be discrete in time but continuous in state-space. Similarly, the queuing model may be a case in continuous time and discrete space by allowing the expected arrival time of the next particle or group of particles to vary continuously.

Consider now a model describing the discharge of a river. An autographic stage recording of the type shown in fig. 1.2 indicates that both time and state-space need to be considered continuously. This type of record can be regarded as a realization of a stationary stochastic process after removal of seasonality and trend. From chapter 4, where the methods of analysing temporal data were discussed, we recognize the autoregressive process as one in which a natural process which is continuous in time and space is treated as a discrete situation by analysing only at fixed periods, e.g. a minute, day or week. By analogy with our discussion of the observation of speed in the previous chapter, we expect that as the time or state-space scale is considered to be finer and finer we have to move, again, into the realms of calculus, and the processes are described by differential equations in time or space.

Imagine the thread of water again deviating from its previous position. If instead of moving one discrete unit, Δ, in one unit of time τ, we allow both Δ and τ, the amount of movement and time taken for it, to become smaller and smaller until they approach zero, then the process becomes continuous in time and space. In this case the stochastic process is called a Wiener or Brownian motion process and it forms the basis of some important geomorphological models. This is because from the Wiener process one may derive the one-dimensional equation for diffusion under some external influence such as gravity, and the one-dimensional equation for heat conduction. As with deterministic models, the solution to these equations depends on initial and boundary conditions. The famous physicist Chandrasekhar developed in 1943 the extension of the continuous random walk model into two and three dimensions and this has been followed by Culling (1963) and others in the development of stochastic models of hillside slope development towards steady-state profiles. We shall outline this paper at some length for it is one of the clearest expositions of a difficult topic.

Culling sets out to consider the proposition, attributed to Strahler (1952), that soil creep is the result of random movements of soil particles. The effects of gravity simply assist or resist randomly directed movements which result from thermal expansion and potential effects; surface tension and capillary forces; cohesive and absorptive tendencies; chemical, electrical and magnetic forces and freezing of soil water. Under his system, movement of particles is controlled essentially by the availability of pore spaces. The process is likened to the diffusion of atoms in the solid state, where the directions of displacement are limited by the crystalline lattice structure. The availability of pore space is a function of the packing and shape of the matrix and interstitial fluids but, provided a large volume is chosen, Culling argues that it is reasonable to assume an even distribution of the spaces.

Each particle is regarded, in the absence of external forces, as being displaced in a random direction and the length of displacement is assumed to be normally distributed with median zero and a very small standard deviation. In other words, if we watched a large number of displacements simulated in a Monte Carlo fashion from the normal distribution, the result would look like a sphere with a high concentration near the centre and a progressively lower concentration outwards which could be represented by the Gaussian or normal curve. If the direction of displacement at any instant of time is independent of the direction at any previous period, then the process is a Markov process. Gravity is viewed as increasing the probability of movement in a downward direction, compacting the lower layers and in effect *decreasing* probability of movement in that direction until some structural equilibrium of pore space is reached, i.e. the soil becomes density layered. The frequency distribution of displacements then reverts to the normal.

Obviously for displacement to take place after an equilibrium has been reached new pore space has to be created at the base of the slope. As displaced particles utilize this extra pore space the holes migrate up-slope, to subdivide and disperse. Eventually it reaches the surface, where it is lost.

In the mathematical development of this model, Culling follows Chandrasekhar (1943) by first considering a one-dimensional random walk which is then extended to three dimensions. This leads to an expression for the probability that a particle will be found in a particular location (volume) after a given number of displacements (compare this with the simple random walk, p. 162). This in turn is developed into the three-dimensional diffusivity equation for the concentration of particles, whose diffusion coefficient is related to the length of successive displacements. The practical test of the theory will depend on empirical values for the average displacement of individual particles, an observation which it has so far been impossible to obtain. The other empirical value required for evaluation of the creep model is the average time a particle remains in one place. Without recourse to actual solutions the expression indicates that 'the effect of the various forces within the soil aggregate tending to produce a randomly directed displacement of the particles results in a slow diffusion of particles from any region of higher concentration towards a neighbouring region of lower concentration at a rate proportional to the concentration gradient.' By relating this concentration gradient to height Culling obtains an equation describing height loss which is very similar to those of the previous chapter.

$$\frac{\partial z}{\partial t} = -K\left(\frac{\partial^2 z}{\partial x^2} + \frac{\partial^2 z}{\partial y^2}\right)$$

$\partial Z/\partial t$ is the partial derivative of height with respect to time and x and y the two dimensional axes. K, the diffusion coefficient of soil particles, is estimated to be 0.5468 if the mean square of the length differences in successive time periods is taken as 1 mm and has the dimensions L^2/T. This equation is useful in studying the denudation of soil-covered slopes provided that the movement of soil particles is limited by the available pore space within it. This excludes mass movement in the sense of aggregate failure, and the effects of running water. In the case of the former the above equation is unsatisfactory and a joint stochastic-mechanical model is required.

Once the diffusion model is accepted then the techniques adopted for its solution are virtually identical with those adopted for deterministic modelling, but it is the formulation of the model from a deductive stochastic viewpoint that is of interest here. The boundary problems adopted are comparable to those of deterministic models and basically control the type of slope development which ensues. Indeed it could be said that a whole range of geomorphological problems is related to choice of the appropriate boundary conditions for solution of the equations of diffusion and Culling treats them in precisely this manner. This convergence of the recent mathematical deterministic and probabilistic modelling on the same body of physical theory is both encouraging and disquieting. Encouraging because it brings the basic working methods in line with those of associated disciplines; disquieting because it implies constraints on geomorphological thought to a specific, albeit well-tried, body of physical theory.

Scheidegger adopts the diffusivity equation model to cover the entire landscape by following an analogy between temperature (T) and height (h) and the change in the quantity of heat (Q) and change in the quantity of mass (M), thus

$$T \longleftrightarrow h$$

and $\quad dQ \longleftrightarrow dM$

Again, this analogy leads directly to the diffusivity equation for height and particular well-known solutions exist which, following Scheidegger (1971), are shown in fig. 7.11a for a mountain range.

Entropy and landforms

The ultimate expression of the approach to geomorphology from the point of view of statistical physics has been in the use of thermodynamic principles as a starting point for modelling. This has arisen principally from the work of Strahler (1950), though interest and impetus was given to the idea by Chorley's (1962) lucid exposition, and further research and extension by Leopold and Langbein (1962), Scheidegger (1964) and Yang (1970, 1971). The idea is discussed here because, in particular, the theory of entropy maximization has been pursued in a statistical manner.

The word entropy was introduced by Clausius to describe a transformation that always occurs when a conversion between thermal and mechanical energy takes place. If a physical system changes from one state to another, each defined by a particular combination of pressure, temperature, composition and magnetic field, then under Clausius' definition the entropy change is calculated by dividing each increment of heat addition by the absolute temperature at which the heat addition occurs, i.e.

$$S = \int_{\mathrm{x1}}^{\mathrm{x2}} \left(\frac{dQ}{T}\right)$$

Where dQ = reversible heat addition, T is absolute temperature and x1 and x2 the bounding states.

In irreversible processes, such as heat flow from a hot to a cold body, this quantity, called entropy, increases as time goes on, or it may remain constant but can never decrease. After an irreversible process there is always more entropy than before, this is the Law of Entropy Increase, generally known as the second law of thermodynamics.

In irreversible processes, as time passes the systems become more disordered. If a layer of blue sand is carefully spread over a layer of white sand in a bottle, vigorous shaking will cause the sands to mix. Continued shaking is not likely to cause the blue and white grains to separate into two layers again. In the flow of heat from a hotter to a cooler body the same principle holds. The temperature of a body is related to the average kinetic energy of its molecules. The molecules of the hot body have, on the average, more kinetic energy than those of the cold object. To begin with, therefore, we have a partial separation between two classes of molecules — those of high and those of lower kinetic energy. After the heat flow has occurred the 'organized', 'ordered' structure of two classes is lost. The high energy molecules are mixed in with lower energy molecules and so disorder increases here also. Thus, the second law can also be stated as, 'the total amount of disorder increases in irreversible processes, and never decreases'.

This disorder can also be expressed in probability terms. If we tossed up a large number of pennies, or a single penny a large number of times, the most likely result would be an equal number of heads and tails. It *is* possible that they would all be heads, or all tails, but these events, which are both very ordered, are extremely unlikely. For most systems a disordered state is more probable. In the same way, after shaking the sand grains, a result in which all the blue sand grains are at the top again is possible but extremely improbable. In other words, in irreversible processes the system changes to a more probable condition (or state).

If we have a measure of the probability of any particular state of the system, we could examine all the states to find out which is the most probable: we could also use it as a measure of the entropy of the system. Shannon developed such a measure in probabilistic terms which is formulated as follows:

$$S = -K\Sigma p_i \log \text{nat}\, p_i$$

in which S is the entropy of the system. The value of S is at a maximum when the probability of each possible state is equal, i.e. when all the (n) states have a probability $p = 1/n$, and at a minimum when only one state is possible and all other states have zero probability.

This entropy concept has become an important element in research methodology, because we can impose barriers which prevent us from reaching the most probable state. Such barriers are constraints on the irreversible process. The research strategy argues as follows: the most probable state is the one in which entropy is maximized; it is also the one which is most disordered and the one for which we need least information. If we obtain the most probable state for any system, from probabilistic reasoning, we have had to make no assumptions about how the system operates. If we then impose a constraint on this most probable situation and examine the results we see only the effects of this constraint.

In other words, we start from the opposite end to more conventional techniques. Instead of creating a very complex model of reality with a very large number of variables, relationships and assumptions, we say let us make no assumptions, obtain the most probable state, then look at the effects of introducing some particular constraint. In mathematical terms, the following steps are involved:

1. define the states;
2. obtain an expression for entropy (as above) subject to a set of constraints expressed quantitatively;
3. obtain the derivative of the expression and set it equal to zero;
4. obtain a solution to the differential equation;
5. examine the resulting configuration of the possible states to see what effect the constraints have had;
6. remodel the system with alternative or additional constraints.

This technique has been used with considerable success by Wilson (1970) for examining characteristics of urban systems. It has been relatively little used in geomorphology partly because of the difficulty of defining the states in such a way that they are compatible with the technique, partly because of the discrete character of the states and partly because of the problem of obtaining solutions to the differential equations which are meaningful in terms of the problem.

An example of the application of the entropy-maximizing technique is von Schelling's modelling of meanders. Here a random walk is imagined in which the frequency distribution of first differences is specified. This means that the angular difference between two steps taken by the channel is drawn from a probability distribution. The most probable path is that for which the entropy of the system, subject to constraints, is most probable. If the entropy of the system without constraints is Q then further constraints F and G may be added. The constraints here are such that the path must pass through a given point $x_j y_j$ after the jth step. In other words, it models the most probable path between two points $x_0 y_0$ and $x_j y_j$. The combined expression $H = Q + \lambda F + uG$, where λ and u are Lagrangian multipliers provided to facilitate solution, is to be maximized since it represents the entropy of the system. This is differentiated and set to zero,simultaneously solving for u and λ and obtaining a set of meander patterns according to the initially specified distributions. If the angular differences are assumed to be drawn from a normal distribution, then the result is shown to be a series of very regular curves. As Scheidegger (1970) points out in discussing the work, what is required is the production of an average (or expected) meander spectrum based on a statistical theory rather than a most probable.

What is of interest here is not so much the actual example or even the technique, but rather the overall problem-solving strategy which is different in character. Its treatment here has been introductory and the reader is referred to the texts by Brillouin (1956) and Wilson (1970) for a fuller discussion.

The second law of thermodynamics and its implications have, of course, been employed in deterministic reasoning in geomorphology, usually for obtaining steady state solutions in dynamic systems and Yang's (1970, 1971) work is exemplary in this respect. In both papers Yang used minimization, subject to constraints, to obtain steady state geometrics for river systems.

As with many theoretical models, realistic evaluation of the outcomes is frequently difficult because real landscapes are very 'noisy' affairs. This is so because of problems of verifying stochastic processes and, in this particular set of models, because steady-state conditions are, in the natural order of things, less common than Hack's (1960) model of landscape development would have us believe. It is in the medium and long time scale that the difficulties are most acute, for it is here that climatic and tectonic instability creates disturbances to the steady state conditions. It seems as though what is needed are more investigations of transient behaviour in theoretical models and less emphasis on steady-state, equilibrium and final-state (entropy maximizing) models.

9 Space and time

Introduction

In the real world events exist in space as well as time; we cannot really consider point locations through time independently of *adjacent* point locations. If events were isolated in space they could easily be handled by the techniques described above. Unfortunately any inferences that could be derived about their performance through time could *not* be extrapolated to adjacent points. Imagine, for example, the difficulty of describing rainfall in a basin where the amount of rain received at each point was independent of that at all adjacent points! As it is, the rainfall at one point has similarities with that at another point; the closer the stations the more similar the precipitation records are.

There is another more important reason for avoiding an artificial decomposition of space and time. This is that area and distance are characteristics which play a crucial role in process in many ways; for example as collecting areas for rainfall, or as controls of momentum and friction. In this sense they play a different and much more explicit role than the passage of time.

A further important difference to consider is the multidirectional character of space. Time goes forward and what has passed is lost in the sense that the system cannot alter *past* values in its feedback controls. In a similar way a geomorphological system cannot anticipate the state of the system ahead of the present time. Space, however, is multi-dimensional: values to the east may be as important as those to the west; channel characteristics ahead of our flood wave can be specified as well as those behind it.

The spatial characteristics of a system are the values taken on by variables as location varies. Sometimes it might be useful to think in terms of frequency-space, for example the wavelengths over which a process is effective. Usually, however, geomorphological problems are described in the conventional Cartesian coordinates, that is by location in a cube-square defined by x, y and z axes. The origin (o, o, o, see fig. 9.1) is arbitrary with respect to any system under consideration. Occasionally there are 'logical' places for the origin, for example at the point of greatest ice thickness in a spatial model of post-glacial uplift. Mathematically the spatial variables can be treated in exactly the same way as other variables such as soil moisture or grain size and so may be manipulated in deterministic models.

It is only possible to add a single further dimension to the two spatial dimensions before losing the capacity to express the system graphically. A basic problem in geomorphology, as in the other sciences, is to represent space, time and several other variables together. Furthermore, in designing geomorphological experiments involving data collection the specific spatial-temporal mix has to be clear in the experimental design. It would be impossible to cover here the whole range of problems of data collection, data analysis and model-building that the introduction of an explicit spatial component involves. Instead, the general typology of space-time models will be outlined as a guide to identification of the main areas of emphasis in mixed space-time problems. Later we give some consideration to the question of scale.

Basic space-time structures

We have chosen the word 'structure' to cover a wide range of procedural mechanisms ranging from data collection designs to stochastic models (table 9.1). Some key characteristics of the structures are such that the type of structure under consideration is relevant whether the activity is model building, empirical searching or data collection.

Zero order

The simplest space time structures are those in which either time or space are considered separately, each independent of the other. Many examples of the former have been discussed already. Downstream variations in channel width, the present location of erosional remnants, the position of an ice margin at a particular time, are but a few examples of the latter. Both types avoid considering space and time together.

Data usually takes the form of a map or a series of values along a line, such as a slope profile. Empirical analysis of the spatial data may take the form of point process analysis, e.g. of potholes, or analysis of line spatial data by methods comparable to those prevalent for time series (ch. 4). Deterministic modelling for temporal structures is relatively easy, even in the absence of spatial data, for example in the analysis of thermal changes at a point in the soil. In cases where spatial variables are given for a single time, deterministic models are not meaningful except in that very special class of models which solve for an 'equilibrium' form that is independent of time. Such is the case for example in the solution of flow-equations to yield a steady-state distribution of soil moisture. Similarly, we find a basic inadequacy in some slope models which solve to a 'characteristic form' but do not yield the intervening stages.

Table 9.1 Classification of types of model on the basis of space-time structures

Order	*Elements*	*Data*	*Empirical*	*Deterministic*	*Probabilistic*
0	Time	Serial time	Autocorrelation at-a-station	Heat conduction in soils	Box-Jenkins models
0	Space	Serial space	and downstream changes in width	Characteristic forms in time-dependent equilibrium	Box-Jenkins models
1	Time discrete + Space discrete	Serial time for several locations	Weekly observations of run-off at several different stations	Models with different parameters at different locations e.g. hydrological models	Markov models for different locations e.g. for thunderstorm patterns
2	Time discrete Space continuous	Spatial form at time intervals	Changing locational properties through time	Change in form properties over time involving thresholds	Random walk models in discrete time
2	Space discrete Time continuous	Continuous stage records at several stations	Autocorrelative properties compared for different locations	Physical sediment yield models with parameter space varying	Regionalized continuous time models
3	Space continuous Time continuous	Virtually none yet obtained	All models discrete	Wave models and hillslope stability models	Diffusion models based on random walks
4	Multi-variate Space-time	Vigil network	Factor analysis of basin or gridded data for several periods	Most hydrological models and models of sediment yield, Luke's hillslope models	Culling's models of probabilistic processes in hillslope development. Simulation models

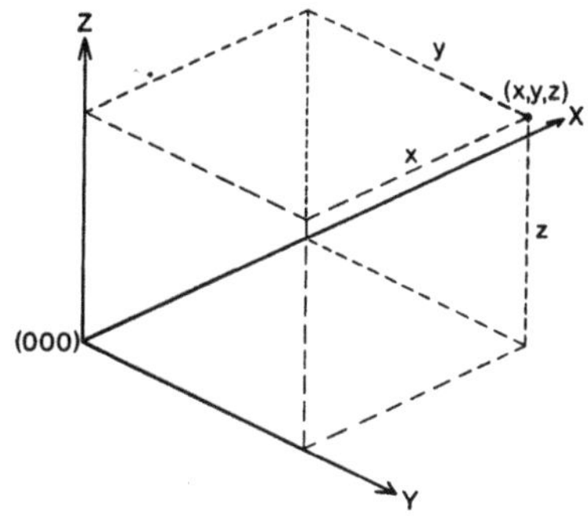

9.1 The Cartesian coordinate system used in conventional 3-dimensional notation.

The same is not necessarily true of probabilistic models for spatial data, for it is possible to 'generate' spatial distributions independently of time, by using the existing configurations. It may occasionally be useful to do so, for example where probabilistic processes yield information about the real processes but unfortunately this is rarely the case (Gudgin and Thornes 1974). As with temporal models they have some use for forecasting, but forecasting spatial phenomena from stochastic models is much more prone to error.

Several studies have attempted to make a case for adopting the statistical notion of ergodicity which was mentioned briefly in chapter 1. This essentially allows for the distribution of a single property at one cross-section in time to be the same as that occurring through time. It has come to mean, in geomorphology, that the *temporal* stages will be present in the same proportions as in a *spatial* distribution for a given time. Unfortunately, the concept of a distributional property has been replaced by one relating to specific sequential development. The identification of a supposed sequence in the forms to be found in the present landscape is fraught with difficulties. Where such a sequence is *known* to exist from dateable fragments (e.g. Brunsden and Kesel 1973) there is no need to invoke the ergodic hypothesis; where a sequence cannot be demonstrated adequately, there is a temptation to put the cases in a supposed sequence, in which case the theory is of no help. This relatively sophisticated abstract concept should be used with care.

First order structures

When both space and time are considered as discrete and only one variable is involved a typical question of interest is the value taken on by the variable at several locations when compared at different observation times. This implies a more rigorous design for data collection than that for zero order models. To be useful it must imply comparability of several locations at the same time or the same location at different times. The locations are discrete. They could, for example, be a set of glacier snouts observed on 1 July every year for 10 different locations in northern Europe. The independent variable might be percentage change in snout position with reference to some local datum.

One might specifically anticipate that empirical analysis would involve cross-spectral investigation if the record were long enough. If the times of observation were not precisely fixed a suitable empirical model could be analysis of variance, in which the contribution made to variance by location was compared with that made by the difference in period. This would be less rigorous in the statistical assumptions that have to be made.

The main relationship of first order structures to deterministic model-building is essentially to provide regional values to the parameters. This means, for example, that a basic model relating water velocity to flow depth, slope and roughness will require a different roughness value for hillslopes as opposed to channels. This regionalization or spatial discreteness provides serious problems in deterministic models and is often a reason for poor replicability. It is frequently overcome by simulating the mathematical model or by allowing the system to change when spatial thresholds are passed. The procedure whereby the submodels are linked is known as coupling. The difficulty of achieving this can perhaps be demonstrated if we ask the reader the following rhetorical question: 'What are the relationships between the magnitude and frequency of sediment movement in a stream channel, the rate of erosion by that stream at the base of its valley-side slope, and the spatial and temporal variability of soil creep on that slope away from the channel?'.

Several important stochastic models are especially appropriate to the discrete time and space character of the first order models. Notable are the auto-regressive models with regionalized parameters. These, as with most other time models, require a regular spacing of the data and require fairly rigorous assumptions about its character. As an example of stochastic modelling in discrete time and space Thornes (1972) showed how the auto-regressive properties of discrete slope series varied according to the time over which the slopes have been moving towards stability.

Second order structures

When one or other of the variables is regarded as continuous the model is second order. The simplest data forms are sequential profiles and fig. 9.2 gives a typical example of the type of information obtained and used. Continuous spatial distributions include the line data so often exhibited on maps such as stream patterns or shorelines. Each time slice may be represented by a map (fig. 9.3). Discrete space-continuous time models are relatively rare in geomorphology. The basic reason for this is that (a) continuous recording of most parameters is technically difficult and relatively expensive, and (b) where such data does exist — for example from tensiometers or thermistors — the analytical procedures, now well known in other sciences for analogue processing, are unknown or relatively little used. The usual procedure is to sample the data and then use the digital output from this scanning. This procedure of breaking up the continuous data into manageable packages (cf. ch. 1) then transforms the problem into a

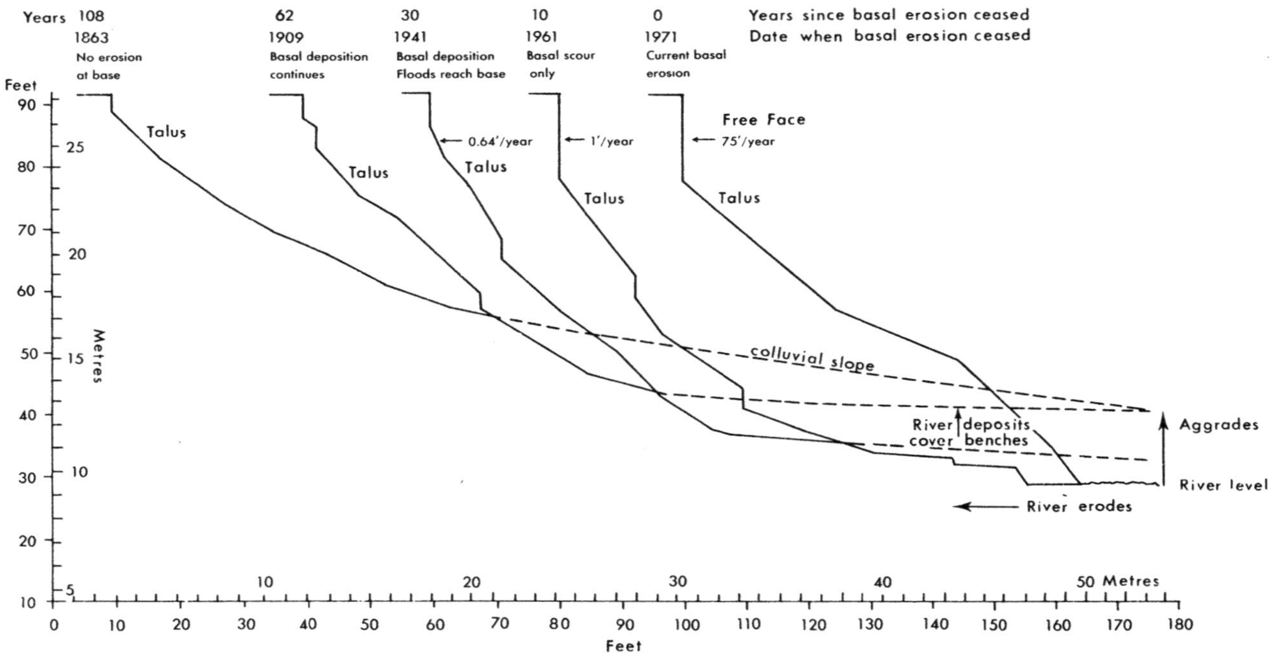

9.2 Sequential slope profiles for a Mississippi river bluff near Port Hudson, Louisiana, illustrating continuous spatial distributions of elevation in which each line represents a 'time slice'. These data constitute a second order space-time structure (after Brunsden and Kesel 1973).

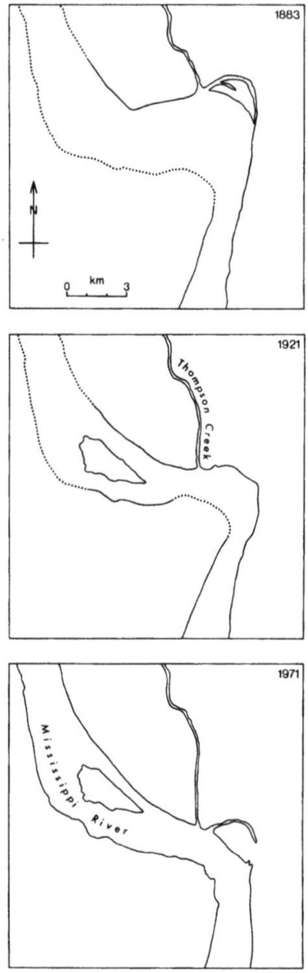

9.3 Three maps which represent successive time slices and show spatial changes in the position of the Mississippi river for 1883, 1921 and 1971. Source: U.S. Geological Survey.

type one structure. This general problem is true of empirical analysis also; no matter whether it is the time or space domain which is continuous it has to be broken up into recognizable patches, which can then be treated analytically as discrete data.

The same is not true of deterministic or stochastic modelling of the data where the opportunity exists for working in the continuous time or space domain. An example of a continuous time, discrete space model is the theoretical model describing surface uplift in response to isostatic adjustments following glaciation. Such models also occur with regionalization of the deterministic model parameters. In the same way stochastic models are given regionalized or time-period parameters which effectively 'package' the space or time components. The well-known Gaussian most frequent random walk generation of meander geometry of von Schelling (1951) is dependent on the value of the random steps which, in a continuous time model, would need to be specified externally by some process.

Third order structures

In the real world both key dimensions are continuous though no way has yet been found of monitoring the data in a continuous fashion. The nearest approach is to replicate a whole landscape block in the laboratory but even then we are not able to capture the results continuously.

Several important models have defined continuous processes and the best we can do is to stop them occasionally for observation. Thus for example models describing the progress of a kinematic wave in a channel, the migration of a knick-point or the development of a surface, are of this character.

If we consider a particle of ice and trace it down a glacier (i.e. keeping space to but one dimension) then the continuous time-space trace is given by the 'characteristic' lines shown in fig. 7.12b, taken from the work of Nye. Bloom (1969) showed that the position of sea-land contact through time could be represented by a three-axis system in which the x-axis represents

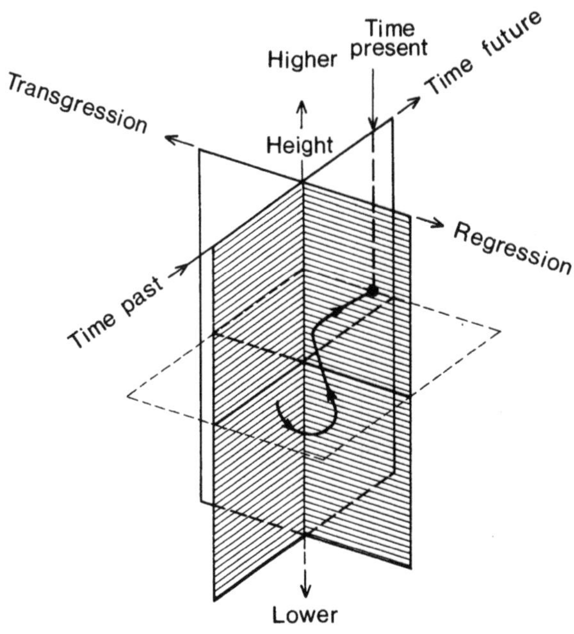

9.4 The passage of a point (sea-level, transverse to the coast) through space and time. Modified from Bloom (1969).

transgression and regression, the y-direction height and the third axis time (fig. 9.4). In applying the model we have to envisage the x-y plane extending at right angles to a point on the coast. In the example the sea-level first falls, regresses, then rises again and transgresses through its present level (represented by $x=0$, $y=0$, $z=0$) toward a higher, transgressive spatial position.

This representation serves to introduce the notion of a space-time cube, in which the three axes represent two spatial directions and time. It represents a sort of continuous time map. A single point in the cube represents a space-time point. A line in the cube must be directed towards the rear of the cube insofar as reversal in time is not allowed. Such a line represents the above mentioned case; the migration of a parcel of water or a flood wave in a channel system, or the movement of an ice crystal inside a glacier. A surface inside the cube is called a manifold and in fig. 9.5 it represents the time-space paths of a meander which is both shifting in position and developing in size. In such a diagram it is only possible to record presence or absence information, i.e. at each location the value taken on by the variable in question is not recorded.

As with second order models all empirical data has to be made discrete, but again mathematical and stochastic models in space and time do not present this difficulty.

Finally, we designate *fourth order structures* as those in which the information concerning several variables, which take on real rather than presence-or-absence form, is concerned. If we use the Berry (1968) data cube for this depiction we find that cells in the cube face represent locations on the one axis and variables on the other. Locations are regarded as discrete, so that each variable represents a specific location (fig. 9.6) whilst the variables may or may not be independent. If we think only of the front face, lines across it represent single variables changing across the various spaces. As an example, consider the observation of drainage density in several specified morphogenetic regions: each morphogenetic region is the location (spaces) coordinate, the variable one of the many possible descriptions of these units.

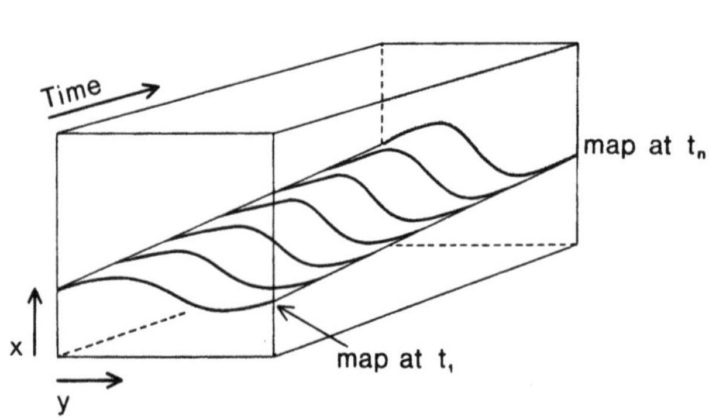

9.5 Manifold in the space-time cube, with space in two dimensions, showing the hypothetical development of a meander.

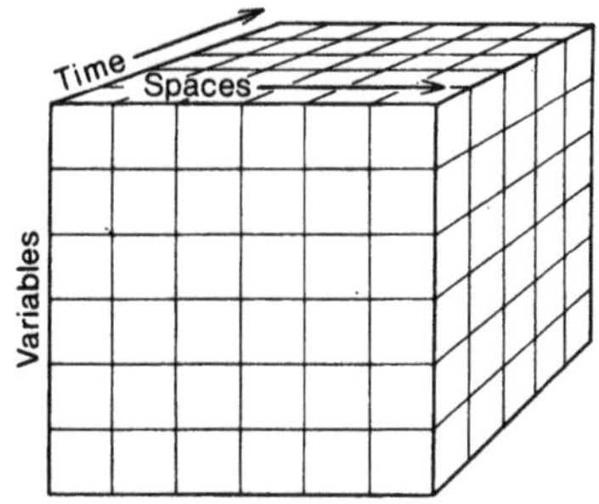

9.6 The space-time cube representing geographical data matrices; developed after Berry (1968).

Similarly we may observe for a single region the values of many variables, as attempted in the regional geomorphological descriptions made in the early- and mid-sixties.

More usefully, by adding time, again as the third axis, we return to the 'real world' situations. Our horizontal planes represent single variables across time and space, as in the observation of global sediment yields; our vertical planes represent multiple variables across time for a single space, as in the complex models of changing drainage basin parameters through time as the effect of the output of sediment yield. In the latter, the drainage basin is seen as a one dimensional artifact in which the variables (rainfall, relief, etc) interact to produce temporal changes. Budget models may often take this form, as for example in the 'idealized' glacier.

Involving all dimensions (one temporal, three spatial and many variables) produces models which attempt to recreate reality in the widest context and so endeavour to obtain a full *explanation* of the real world situation. It is to this end that much modern geomorphology is directed.

Scale

To attempt this requires consideration of the question 'To what extent is scale matching between time and space pre-ordained or at least required by a particular modelling technique?' Intuitively one might expect that there is a link between time and space through the operation of rate. There is such a link of course, but there is no reason to suppose that it imposes rigid matching constraints on our modelling. Indeed it is often convenient, if not terribly realistic, to avoid the process. Sometimes conditions force us to do this because of limitations on our ability to collect and handle data.

By *scale* we refer to the observation that most processes produce variability in areal or temporal series which is best identified at particular intervals of observation rather than continuously. If we are observing a spatial or temporal series, different processes produce variability at different scales. By and large, however, we have little understanding of this allocation of sources of variance, by process, across the different scale levels.

We have two kinds of problem to cope with:

(i) identification of the process producing the variance at a particular scale level or levels, and

(ii) knowing the processes, to determine at what scale level they are most likely to be effective.

In the first case a crude example is storm-runoff and hillslope development. Here we are identifying short and long time scales and need to match the process modelling accordingly; for example, we would not anticipate changes in hillslope geometry in modelling storm run-off. As with statistical modelling, so with mathematical models it is generally accepted that while one can always aggregate upward, it is not possible to disaggregate if the data net has too wide a mesh in the first place. This often becomes an argument for inefficiency in investigation.

In the second case, although we are aware of a process, capturing the effects of the process at the appropriate level is not always easy. Thus, for example, we do not know the desirable scale for measurement for many relatively slow processes, yet both time and space scales have to be correct if we are to make a full descriptive statement.

It is generally true to say that a slow process involving movement requires a long time to cover a large area, and must be observed over a long rather than a short time if the major characteristics of the process are to be evaluated. Thus observations of the minute movements of the glacier snout are not likely to reveal the large and long changes accompanying continental glaciation. Similarly, a rapid process involving movement will quickly cover a given area and its observation will require a high density of observation points in time and space. However, we have to conclude that since the *scale* at which we observe is an artifact of convenience there is no intrinsic reason why *scales* of observation should be the same for time and space. This is not of course saying that time and space are not inextricably mixed but that this is at present a philosophical rather than a practical question.

In practical terms, the resolution is determined, at least empirically, by the cost (however measured) and difficulty of obtaining information. Theoretical studies are not prescribed by these limits, so that Davis was able to sketch the development of continents over very long time periods and Scheidegger is able to conceive of the diffusion of hillslope mass towards an entropy maximizing state. It is with the validation of these theoretical models or the development of empirical ones that the constraints are imposed.

If one considers the broad spectrum of models we observe that there is a shift in the time and space scales according to the prevailing fashions. At the beginning of the century the studies centred mainly on two types, the emphasis on large scale, long term denudation chronologies, and long term but essentially local studies of glacial chronology. Detailed micro-studies of process in the laboratory flume were relatively few in number. Towards the middle of the century there had been a perceptible shift towards medium space and low time scales of observation as sub-regional denudation chronologies became the vogue. In the sixties we saw a shift towards the short-time and local space-scale studies with the detailed process-type of observation, while minor peaks occur representing glacial-chronological studies. At present in the mid-seventies, the shift appears to be towards the medium time and space scales with models of hillslope development in a catchment framework.

Finally, we observe that there must be a strong relationship between space and time scales and the magnitude and frequency of the process. Processes of low spatial and temporal frequencies must be observed on the corresponding scales. Unfortunately, low frequency events of high magnitude cause very great difficulties, for often the magnitude is sustained for a short time period over a very considerable area and the scales are here at odds with one another. In the same way extremely slow processes are exceedingly difficult to cope with and hence the attempts to solve them by using the space-time substitutions. One of the more appealing features of the contemporary fashion for slope and sediment yield studies in medium-sized catchments (of the order of tens of square kilometres) is the ability to grasp and measure the events in terms of our own particular time scales. This changing fashion has the result that we are beginning to know much about processes in 'man-sized' areas over 'human' lifespans. We continue to give much theoretical consideration to landform evolution over 'cyclic' time, but the challenge remains of extrapolating the short term record of processes to the relatively unknown time span of say 100-10,000 years and beyond.

Bibliography

AHNERT, F. (1970) Functional relationships between denudation relief and uplift in large, mid-latitude drainage basins. *Am. J. Sci.* 268, 243–63.

ANDERSON, H. W. (1971) Recovery of suspended sediment discharge accelerated by major floods and poor land use. *Fall Meeting Am. Geophys. Un.* Abstract EOS, TAGU, 52 (ii), 829.

ANDERSON, J. L. and SOLLID, J. L. (1971) Glacial chronology and glacial geomorphology in the marginal zones of the glaciers, Midtdalsbreen and Nigardsbreen, south Norway. *Norsk. Geogr. Tidsch.* 25 (1), 1–38.

ANDREWS, J. T. (1970) *A geomorphological study of post-glacial uplift, with particular reference to Arctic Canada.* Inst. Brit. Geogr. Spec. Publ. 2. 156.

ANDREWS, J. T. and WEBBER, P. J. (1964) A lichenometrical study of the northwestern margin of the Barnes Ice Cap: a geomorphological technique. *Geogr. Bull.* 22, 80–104.

ANDREWS, J. T. and WEBBER, P. J. (1969) Lichenometry to evaluate changes in glacial mass budgets. *Arctic and Alpine Res.* 1, 181.

ANTEVS, E. (1953) Geochronology of the deglacial and Neothermal ages. *J. Geol.* 61, No. 3, 195–230.

ANTEVS, E. (1955) Geologic-climatic dating in the west. *Am. Antiq.* 20 (4), Pt 1, 317–35.

ARBER, M. A. (1940) The coastal landslips of south-east Devon. *Proc. Geol. Assoc.* 51 (3), 257–71.

BAGNOLD, R. A. (1966) An approach to the sediment transport problem from general physics. *US Geol. Surv. Prof. Pap.* 422–I, 37.

BAKKER, J. P. and LE HEUX, J. W. N. (1952) A remarkable new geomorphological law. *Koninklijke Nederlandsche Akademie van Wetenschappen.* 55 (B), 399–410, 554–71.

BENEDICT, J. B. (1967) Recent glacial history of an alpine area in the Colorado Front Range, USA. *J. Glac.* 6 (48), 817–32.

BERRY, B. J. L. (1968) Approaches to regional analysis: a synthesis. *Ann. Assoc. Am. Geogr.* 54, 2–11.

BESCHEL, R. E. (1961) Dating rock surfaces by lichen growth and its application to glaciology and physiography (lichenometry). In RAASCH, G. O. (ed.), *Geology of the Arctic.* Proc. Ist Int. Symp. on Arctic Geology, Calgary, Alberta 11-13 Jan. 1960. Toronto. 2, 1044–62.

BIRMAN, J. H. (1964) Glacial geology across the crest of the Sierra Nevada, California. *Geol. Soc. Am. Spec. Pap.* 75, 80.

BIROT, P. (1960) Le cycle d'érosion sous les differents climats. *Curso de Altos Estudos Geograficos*. Centre des Pesquisas de Geografico do Brazil. 1.
BLOOM, A. L. (1965) The explanatory description of coasts. *Zeitschr. Geom.* 9, 422–36.
BLOOM, A. L. (1969a) Holocene submergence in Micronesia as the standard for eustatic sea-level changes. Lecture (multigraphed) *Symposium on the evolution of shorelines and continental shelves during the Quaternary*. UNESCO, Paris.
BLOOM, A. L. (1969b) *The surface of the earth*. New Jersey.
BOER, G. de (1964) Spurn Head: its history and evolution. *Trans. Inst. Brit. Geogr.* 34, 71–89.
BOLTON, G. S. and WORSLEY, P. (1968) Late Weichselian glaciation of the Cheshire–Shropshire Basin. *Nature*, 207, 704–6.
BORNFELDT, F. and ÖSTERBORG, M. (1958) Lavarter som hjälpmedel i datering av ändmoräner vid norska glaciärer. *Stockholm Högskola, Geografiska Proseminariet*. October (microfilm).
BRILLOUIN, L. (1956) *Science and information theory*. New York (2nd ed. 1962).
BROWN, E. H. (1960) *The relief and drainage of Wales*. Cardiff.
BROWN, E. H. (1961) Britain and Appalachia: a study in the correlation and dating of planation surfaces. *Trans. Inst. Brit. Geogr.* 29, 91–100.
BROWN, E. H. (1970) Man shapes the earth. *Geogr. J.* 136, 73–85.
BROWN, J. C. (1962) The drainage pattern of the lower Ottawa valley. *Can. Geog.* 6 (1), 22–31.
BRUNSDEN, D. (1973) The application of system theory to the study of mass movement. *Geologia Applicata e Idrogeologia*. Bari. 7 (1), 185–207.
BRUNSDEN, D. (1974) The degradation of a coastal slope, Dorset. In BROWN, E. H. and WATERS, R. S. (eds) *Progress in geomorphology*, papers in honour of David L. Linton. *Inst. Brit. Geogr. Spec. Publ.* 7, 79–98.
BRUNSDEN, D. and JONES, D. K. C. (1972) The morphology of degraded landslide slopes in south-west Dorset. *Q. J. Eng. Geol.* 5 (3), 205–22.
BRUNSDEN, D. and JONES, D. K. C. (1976) The evolution of landslide slopes. *Phil. Trans. Royal Soc.*
BRUNSDEN, D. and KESEL, R. H. (1973) The evolution of a Mississippi river bluff in historic time. *J. Geol.* 81, 576–97.
BRYAN, K. (1923) Erosion and sedimentation in the Papago County, Arizona. *US Geol. Surv. Bull.* 730, 19–90.
BRYAN, K. (1940a) The retreat of slopes. *Ann. Assoc. Am. Geogr.* 30, 254–67.
BRYAN, K. (1940b) Gully gravure: a method of slope retreat. *J. Geomorph.* 3, 87–107.
BRYAN, K. (1946) Cryopedology — the study of frozen ground and intense frost action with suggestions of nomenclature. *Am. J. Sci.* 244, 622–42.
BUCKLAND, W. (1840) On the landslipping near Axmouth. *Proc. Ashmolean Soc.* 1, 1832–42.
BUTZER, K. W. (1964) *Environment and archeology*. London.
CALVER, A., KIRKBY, M. J. and WEYMAN, D. R. (1972) Modelling hillslope and channel flows. In CHORLEY, R. J. *Spatial analysis and geomorphology*. London.
CARR, A. P. (1962) Cartographic records and historical accuracy. *Geogr.* 47 (2), 135–44.
CARSON, M. A. and KIRKBY, M. J. (1972) *Hillslope form and process*. Cambridge.
CARSON, M. A. and PETLEY, D. J. (1970) The existence of threshold hillslopes in the denudation of the landscape. *Trans. Inst. Brit. Geogr.* 49, 71–95.

CARTER, C. S. and CHORLEY, R. J. (1961) Early slope development in an expanding stream system. *Geol. Mag.* 98, 117–30.

CHAMBERLIN, T. C. (1883) Terminal moraine of the Second Glacial Epoch. *US Geol. Surv. 3rd Ann. Rep.* 1881-2, (1883), 291–402.

CHANDRASEKHAR, S. (1943) Stochastic problems in physics and astronomy. *Rev. Mod. Physics.* 15 (1), 1–89.

CHORLEY, R. J. (1962) Geomorphology and general systems theory. *US Geol. Surv. Prof. Pap.* 500 (B).

CHORLEY, R. J. (1964) Geography and analogue theory. *Ann. Assoc. Am. Geogr.* 54, 127–37.

CHORLEY, R. J. (1965) A re-evaluation of the geomorphic system of W. M. Davis. In CHORLEY, R. J. and HAGGETT, P. (eds) *Frontiers in geographical teaching.* London. Ch. 2, 21–38.

CHORLEY, R. J. (1967) Models in geomorphology. In CHORLEY, R. J. and HAGGETT, P. (eds) *Models in geography.* London. 59–96.

CHORLEY, R. J. and KENNEDY, B. (1971) *Physical geography: A systems approach.* London. 370.

CHORLEY, R. J., DUNN, A. J. and BECKINSALE, R. P. (1964, 1973) *The history of the study of landforms.* London. Vol. 1 *Geomorphology before Davis,* Vol. 2 *The life and work of William Morris Davis.*

CHURCHWARD, H. M. (1961) Soil studies at Swan Hill, Victoria, Australia. I. Soil layering. *J. Soil Sci.* 12 (1), 73–86.

CLAYTON, K. M. (1957) Some aspects of the glacial deposits of Essex. *Proc. Geol. Assoc.* 68, 1–21.

COLBY, B. R. (1964) Scour and fill in sand-bed streams. *US Geol. Surv. Prof. Pap.* 462 (D), 32.

CONYBEARE, W. D. (1840) Extraordinary landslip and great convulsions of the coast of Culverhole Point, near Axmouth. *New Phil. J.* 29.

CONYBEARE, W. D. *et al.* (1840) *Ten plates comprising a plan, sections and views representing the changes produced on the coast of east Devon between Axmouth and Lyme Regis by the subsidence of the land and the elevation of the bottom of the sea, on 26th December 1839 and 3rd February 1840.* London.

COOK, F. A. and RAICHE, V. G. (1962) Freeze-thaw cycles at Resolute, N.W.T. *Geol. Bull.* 18, 64–78.

COOPER, H. H. Jnr and RORABAUGH, M. I. (1963) Ground water movements and bank storage due to flood stages in surface streams. *US Geol. Surv. Water Supply Pap.* 1536-J.

CORBEL, J. (1957) L'érosion climatique des granites et silicates sous climates chauds. *Rev. Géom. Dyn.* 8, 4–8.

CORBEL, J. (1959) Vitesse de l'érosion. *Zeits. für Geom.* 3, 1–28.

CORBEL, J. (1964) L'érosion terrestre, étude quantitative (methodes — techniques — résultats). *Ann. Géograph.* 73, 385–412.

CULLING, W. E. H. (1960) Analytical theory of erosion. *J. Geol.* 69, 336–44.

CULLING, W. E. H. (1963) Soil creep and the development of hillslide slopes. *J. Geol.* 71, 127–61.

CULLING, W. E. H. (1965) Theory of erosion on soil covered slopes. *J. Geol.* 73, 230–54.

DARWIN, C. R. (1839) *Narrative of the surveying voyages of H.M.S. 'Adventure' and 'Beagle' between 1826 and 1836.* Vol. I-III. London.

DARWIN, C. R. (1859) *On the origin of species by means of natural selection; or, the preservation of favoured races in the struggle for life.* London.

DAVIS, M. B. and GOODLETT, J. C. (1960) Comparison of the present vegetation

and pollen spectra in surface samples from Browningtn Pond, Vermont. *Ecol.* 41, 346–57.
DAVIS, W. M. (1899) The geographical cycle. *Geogr. J.* 14, 481–504.
DAVIS, W. M. (1905) Complications of the geographical cycle. *Rep. 8th Int. Geogr. Congress Washington, 1904.* 150–63.
DAVIS, W. M. (1905) The geographical cycle in an arid climate. *J. Geol.* 13, 381–407.
DAVIS, W. M. (1906) The sculpture of mountains by glaciers. *Scott. Geogr. Mag.* 22, 76–89. *Abst. Rep. Brit. Assoc. Adv. Sci.* 75, 393–4.
DAVIS, W. M. (1909) *Geographical essays* (ed. JOHNSON, D. W.). Boston, 777. Republ. 1954, New York.
DAVIS, W. M. (1909) Glacial erosion in North Wales *Q. J. Geol. Soc.* 65, 281–350.
DAVIS, W. M. (1922) Peneplains and the geographical cycle. *Bull. Geol. Soc. Am.* 33, 587–98.
DAVIS, W. M. (1930) Rock floors in arid and in humid climates. *J. Geol.* 38, 1–27, 136–58.
DAVIS, W. M. (1933) Granite domes of the Mojave Desert, California. *Trans. San Diego Soc. Nat. Hist.* 7, 211–58.
DAVIS, W. M. (1936) Geomorphology of mountainous deserts. *Rep. 16th Int. Geol. Congr. 1933.* 2, 703–14.
DAVIS, W. M. (1938) Sheetfloods and streamfloods. *Bull. Geol. Soc. Am.* 49, 1337–416.
DERBYSHIRE, E. (1973) *Climatic geomorphology.* London.
DOLE, R. B. and STABLER, H. (1909) Denudation. *US Geol. Surv. Water Supply Pap.* 234, 78–93.
DUNNE, T. and BLACK, R. D. (1970) An experimental investigation of runoff production in permeable soils. *Water Resources Res.* 6, 478–90.
DOUGLASS, A. E. (1914) A method of estimating rainfall by the growth of trees. In HUNTINGTON, E. *The climatic factor as illustrated in arid America.* Washington. 101–38.
DOUGLASS, A. E. (1946) Precision of ring dating in tree-ring chronologies. *Univ. Arizona Bull.* 17, 1–21.
DURY, G. H. (1964a) Principles of underfit streams. *US Geol. Surv. Prof. Pap.* 452–A, 67.
DURY, G. H. (1964b) Subsurface explorations and chronology of underfit streams. *US Geol. Surv. Prof. Pap.* 452–B, 56.
DURY, G. H. (1965) Theoretical implications of underfit streams. *US Geol. Surv. Prof. Pap.* 452–C, 43.
DUTTON, C. E. (1871–2) The causes of regional elevation and subsidences. *Proc. Am. Phil. Soc.* 12, 70–2.
DUTTON, C. E. (1880–1) The physical geology of the Grand Canyon district. *US Geol. Surv. Ind. Ann. Rep. (1880–1).* 46–166.
DUTTON, C. E. (1889) On some of the greater problems of physical geology. *Bull. Phil. Soc. Washington.* 11, 51–64.
EAGLESON, P. S. (1970) *Dynamic hydrology,* New York.
EDWARDS, A. M. C. and THORNES, J. B. (1973) Annual cycle in river water quality: a time series approach. *Water Resources Res.* 9 (5), 1286–95.
EICHER, D. L. (1968) *Geologic time.* Englewood Cliffs N.J.
ELIAS, F. (1963) Precipitaciones maximas en España. *Min. de Agric. Servicio de Conservacion de Suelos,* Madrid.
EMILIANI, C. (1955) Pleistocene temperatures. *J. Geol.* 63, 538–78.
EMILIANI, C. (1966) Palaeotemperature analysis of Caribbean cores and a generalised temperature curve. *J. Geol.* 74, 109–26.

EMILIANI, C. (1967) The generalised temperature curve for the past 425,000 years. *J. Geol.* 75, 504–10.
EMMETT, W. W. (1970) The hydraulics of overland flow on hillslopes. *US Geol. Surv. Prof. Pap.* 662-A, 68.
ERDTMAN, G. (1954) *An introduction to pollen analysis.* Chronica Botanica (1st ed. 1943). Waltham: Mass. 239.
EVANS, I. S. (1972) General geomorphometry, derivation of altitude and descriptive statistics. In CHORLEY, R. J. (ed.) *Spatial analysis in geomorphology.* London. 17–90.
EVANS, P. (1971) Towards a Pleistocene time-scale. Pt 2 of *The Phanerozoic time scale: A supplement.* Spec. Publ. 5, Geological Society London. 121–356.
FAIRBRIDGE, R. W. (1961) Eustatic changes in sea-level. In *Physics and chemistry of the earth*, 4. Oxford. 99–185.
FISHER, O. (1866) On the disintegration of a chalk cliff. *Geol. Mag.* 3, 354–6.
FLAXMAN, E. M. and HIGH, R. D. (1955) Sedimentation in drainage basins of the Pacific Coast states. *Soil Conserv. Serv.* (mimeographed). 8.
FLEMING, G. (1969) Design curves for suspended load estimation. *Proc. Inst. Civ. Engr.* 43, 1–9.
FLINT, R. F. (1970) *Glacial and Pleistocene geology.* New York (1st edn 1957). 553.
FOURIER, J. B. (1822) *Théorie analytique de la chaleur.* Paris.
FOURNIER, F. (1960) *Climat et érosion: la relation entre l'érosion du sol par l'eau et les precipitations atmosphériques.* Paris. 201.
FOURNIER, F. (1969) Transports solides affectués par les cours d'eau. *Bull. Internat. Ass. Sci. Hydrol.* 14, 7–47.
FREEMAN, J. R. (1922) Flood problems in China. *Trans. Am. Soc. Civ. Engrs.* 85, 1436.
FRITTS, H. C. (1966) Growth rings of trees, their correlation with climate. *Sci.* 154, 973–9.
FRYE, J. C. and LEONARD, A. B. (1957) Ecological interpretation of Pliocene and Pleistocene stratigraphy in the Great Plains region. *Am. J. Sci.* 255, 1–11.
GARDNER, J. (1969) Observation of surficial talus movement. *Zeits. für Geom.* 13, 317–23.
GARNER, H. F. (1968) Climatic geomorphology. In FAIRBRIDGE, R. W. (1968) *Encyclopedia of geomorphology.* New York. 129–30.
GEER, G. de (1912) A geochronology of the last 12,000 years. Compte Rendu *11th Int. Geol. Cong. Stockholm 1910* 1, 241–58.
GEIKIE, A. (1868) On denudation now in progress. *Geol. Mag.* 5, 249–54.
GEIKIE, A. (1873) Earth sculpture and the Huttonian school of geology. *Trans. Edin. Geol. Soc.* 2, 247–67.
GILBERT, G. K. (1877) *Report on the geology of the Henry Mountains.* Washington. 160.
GILBERT, G. K. (1909) The convexity of hilltops. *J. Geol.* 17, 344–50.
GILLULY, J. (1949) Distribution of mountain building in geologic time. *Bull. Geol. Soc. Am.* 60, 561–90.
GLOCK, W. S. (1937) *Principles and methods of tree-ring analysis.* Washington. 486.
GREEN, J. F. N. (1936) The terraces of southernmost England. *Q. J. Geol. Soc.* 92, 58–88.
GREEN, J. F. N. (1943) The age of the raised beaches of southern Britain. *Proc. Geol. Assoc.* 54, 129–40.
GREGORY, K. J. and WALLING, D. E. (1973) *Drainage basin form and process: a geomorphological approach.* London.
GREGORY, K. J. and WALLING, D. E. (eds) (1974) *Fluvial processes in*

instrumented Watershed. Brit. Geomorph. Res. Gp. Inst. Brit. Geog. Spec. Publ. 6.

GRODINS, F. S. (1963) *Control theory in biological systems.* New York/London. 205.

GUDGIN, G. and THORNES, J. B. (1974) Probability in geographic research applications and problems. *The Statistician*, 23, 157–77.

GUMBEL, E. J. (1958) *Statistics of extremes.* New York. 375.

GUPTA, V. L. (1973) Information content of time-variant data. *Proc. Am. Soc. Civ. Engrs. J. Hydraul. Div.* 99, HY3, 383–94.

GUTHRIE-SMITH, H. (1926) *Tutira: the story of a New Zealand sheep station.* Private. Edinburgh and London (3rd edn 1953).

GUTENBERG, B. (1941) Changes in sea-level, postglacial uplift and mobility of the earth's interior. *Bull. Geol. Soc. Am.* 52, 721–72.

HACK, J. T. (1960) Interpretation of erosional topography in humid temperate regions. *Am. J. Sci.* 258A, 80–97.

HACK, J. T. (1965) Geomorphology of the Shenandoah valley, Virginia and West Virginia, and origin of the residual ore deposits. *US Geol. Surv. Prof. Pap.* 484.

HACK, J. T. (1966) Interpretation of the Cumberland escarpment and highland rim, south-central Tennessee and north-east Alabama. *US Geol. Surv. Prof. Pap.* 524–6, 16.

HALL, E. T. (1969) Dating pottery by thermoluminescence. In BROTHWELL, D. and HIGGS, E. *Science in archaeology: a survey of progress and research.* London. 106–8.

HARBAUGH, J. W. and BONHAM-CARTER, G. (1970) *Computer simulation in geology.* New York.

HARVEY, D. (1969) *Explanation in geography.* London.

HEWITT, K. (1967) Ice front deposition and the seasonal effect: a Himalayan example. *Trans. Inst. Brit. Geogr.* 42, 93–106.

HEWITT, K. (1970) Probablistic approaches to discrete natural events. A review and theoretical discussion. *Econ. Geogr.* 46 (2), (suppl), 332–49.

HILTON, P. J. (1968) *Differential calculus.* London.

HIRANO, M. (1968) A mathematical model of slope development. An approach to the analytical theory of erosional topography. *J. of Geosci.* Osaka. 11 (2), 14–52.

HOLEMAN, J. N. (1968) The sediment yield of the major rivers of the world. *Water Resources Res.* 4, 737–47.

HORTON, R. E. (1933) The role of infiltration in the hydrologic cycle. *Trans. Am. Geophys. Un.* 14, 446–60.

HORTON, R. E. (1939) The analysis of runoff plot experiments. *Trans. Am. Geophys. Un.* 20, 693.

HORTON, R. E. (1945) Erosional development of streams and their drainage basins: hydrophysical approach to quantitative morphology. *Bull. Geol. Soc.* 56, 275–370.

HOWARD, A. D. (1965) Geomorphological systems — equilibrium and dynamics *Am. J. Sci.* 263, 302–12.

HOWARD, A. D. (1971a) Simulation model of stream capture. *Bull. Geol. Soc. Am.* 82, 1355–76.

HOWARD, A. D. (1971b) Simulation of stream networks by headward growth and branching. *Geol. Anal.* 3, 29–50.

HUBBELL, D. W. (1964) Apparatus and techniques for measuring bedload. *US Geol. Surv. Water Supply Pap.* 1748.

HURST, H. E. (1950) Long-term storage capacity of reservoirs. *Trans. Am. Soc. Civ.*

Engrs. 116, 770–808.
HUTCHINSON, J. N. (1967) The free degradation of London Clay cliffs. *Proc. Geotech. Conf. Oslo* 1, 113–18.
HUTTON, J. (1795) *Theory of the earth, with proofs and illustrations.* Edinburgh.
JACKLI, H. (1957) Gegenwartsgeologie des bundnerischen Rheingebietes — ein Beitrag zur exogenen Dynamik Alpiner Gerbirgslandschaften. *Beitr. zur Geol. der Schweiz. Geotechn.* Serie 36.
JAMES, P. A. (1971) The measurement of soil frost-heave in the field. *Brit. Geomorph. Res. Gp. Tech. Bull.* 8, 43.
JELGERSMA, S. (1961) Holocene sea-level changes in the Netherlands. *Mededel. Geol. Sticht.* Ser. C 6 (7), 1–101.
JELGERSMA, S. (1966) Sea-level changes during the last 10,000 years. *Proc. Int. Symp. – World climate from 8000 to 0 B.C.* London. 54–71.
JENKINS, G. M. and WATTS, D. G. (1968) *Spectral analysis and its applications* San Francisco.
JENNINGS, M. E. and BENSON, M. A. (1969) Frequency curves for annual flood series with some zero events or incomplete data. *Water Resources Res.* 5 (1), 276–80.
JOCHIMSEN, M. (1966) Ist die grosse das flechtenthallus wirklich ein brauchbarer masstab zur datierung von glazialmorphologischen relikten. *Geogr. Ann.* 48(A), 157.
JOHNSON, R. H. (1965) A study of the Charlesworth landslides near Glossop. *Trans. Inst. Brit. Geogr.* 37, 111–26.
JOHNSON, D. W. (1919) *Shore processes and shoreline development.* New York. 584.
JOHNSON, D. W. (1931) *Stream sculpture on the Atlantic slope: A study of the evolution of the Appalachian rivers.* New York.
JONES, D. K. C. (1970) *The Vale of Brooks: field guide.* Inst. Brit. Geog. Ann. Meeting. 43–6.
JONES, D. K. C. (1974) The influence of the Calabrian transgression on the drainage evolution of south-east England. In BROWN, E. H. and WATERS, R. S. *Progress in geomorphology.* Papers in honour of D. L. Linton. Inst. Brit. Geog. Spec. Pub. 7, 139–58.
JONES, O. T. (1924) The longitudinal profiles of the Upper Towy drainage system. *Q. J. Geol. Soc.* 80, 568–609.
KEEBLE, A. B. (1971) Freeze-thaw cycles and rock weathering in Alberta. *Albertan Geogr.* 7, 34–41.
KELLERHALS, R. and BRAY, D. J. (1971) Sampling procedures for coarse fluvial sediments. *Proc. Am. Soc. Civ. Engrs.* Journ. Hydraulics Div. 97, HY8, 1165-80.
KERNEY, M. P. (1963) Late glacial deposits on the chalk of south-east England. *Phil. Trans. Royal Soc.* Ser. B. Biol. Sci. 730, 246, 203–54.
KERNEY, M. P., BROWN, E. H. and CHANDLER, T. J. (1964) The Late-glacial and Post-glacial history of the chalk escarpment near Brook, Kent. *Phil. Trans. Roy. Soc.* Ser. B, 248, 135–204.
KIDSON, C. (1952) Dawlish Warren: a study of the evolution of the sand spits across the mouth of the river Exe in Devon. *Trans. Inst. Brit. Geogr.* 18, 69–80.
KING, C. A. M. (1969) Glacial geomorphology and chronology of Henry Kater Peninsula, East Baffin Island, N.W.T. *Arctic and Alpine Res.* 1 (3), 195–212.
KING, C. A. M. (1972) *Beaches and coasts* (2nd ed). London.
KING, C. A. M. and McCULLAGH, M. J. (1971) A simulation model of a complex recurved spit. *J. Geol.* 79, 22–37.
KING, L. C. (1953) Canons of landscape evolution. *Bull. Geol. Soc. Am.* 64, 721–62.

KING, L. C. (1955) Pediplanation and isostasy: an example from South Africa. *Q. J. Geol. Soc.* 111, 353–9.

KING, L. C. (1957) The uniformitarian nature of hillslopes. *Trans. Edinburgh Geol. Soc.* 17, 81–102.

KING, L. C. (1962) *Morphology of the earth.* Edinburgh and London.

KIRKBY, M. J. (1971) Hillslope process-response models based on the continuity equation. In BRUNSDEN, D. (comp.) *Slopes: form and process.* Inst. Brit. Geogr. Spec. Pub. 3, 15–30.

KNAPP, B. H. (1973) A system for the field measurement of soil water movement. *Tech. Bull. Brit. Geomorph. Res. Gp.* 9.

KRUMBEIN, W. C. (1966) A comparison of polynomial and Fourier models in map analysis. *Office of Naval Research Project 388-078 Tech. Report 2.* Evanston.

KRUMBEIN, W. C. and GRAYBILL, F. A. (1965) *An introduction to statistical models in geology*, New York.

KRUMBEIN, W. C. and SLOSS, L. L. (1963) *Stratigraphy and sedimentation* (2nd edn). San Francisco.

LARSSON, W. (1937) Vulkanische Asche vom Ausbruch des Chilenishen Vulkans Quizapu (1932) in Argentine gesammelt. *Uppsala. Geol. Inst. Bull.* 26, 27–52.

LAMARCHE, V. C., Jr (1968) Rates of slope degradation as determined from botanical evidence, White Mountains, California. *US Geol. Surv. Prof. Pap.* 352-J, 377.

LANGBEIN, W. B. and LEOPOLD, L. B. (1964) Quasi-equilibrium states in channel morphology. *Am. J. Sci.* 262, 782–94.

LANGBEIN, W. B. and LEOPOLD, L. B. (1966) River meanders: theory of minimum variance. *US Geol. Surv. Prof. Pap.* 422-H.

LANGBEIN, W. B. and SCHUMM, S. A. (1958) Yield of sediment in relation to mean annual precipitation. *Trans. Am. Geophys. Un.* 39, 1076–84.

LEES, G. M. (1955) Recent earth movements in the Middle East. *Geologische Rundschau* 43, 221–6.

LEHMANN, O. (1933) Morphologische theorie der Verwitterung von steinschlagwänden. *Vierteljahrsschrift der Naturforschende Gesellschaft* 87, 83–126.

LEIGHLY, J. (1940) Comments in: Walther Penck's contribution to geomorphology. Symposium (1939), *Ann. Assoc. Am. Geogr.* 30, 219–80.

LEOPOLD, L. B. and DUNNE, T. (1971) Field methods for hillslope description. *Tech. Bull. Brit. Geomorph. Res. Gp.* 7.

LEOPOLD, L. B. and EMMETT, W. W. and MYRICK, R. M. (1966) Channel and hillslope processes in a semi-arid area, New Mexico. *US Geol. Surv. Prof. Pap.* 352-G, 193–253.

LEOPOLD, L. B. and LANGBEIN, W. B. (1962) The concept of entropy in landscape evolution. *US Geol. Surv. Prof. Pap.* 500-A, 3, 20.

LEOPOLD, L. B. and LANGBEIN, W. B. (1963) Association and indeterminancy in geomorphology. In ALBRITTON, C. C. (ed.) *The fabric of geology.* 184–92.

LEOPOLD, L. B. and MADDOCK, T. Jr, (1953) The hydraulic geometry of stream channels and some physiographic implications. *US Geol. Surv. Prof. Pap.* 252, 57.

LEOPOLD, L. B., WOLMAN, M. G. and MILLER, J. P. (1964) *Fluvial processes in geomorphology.* San Francisco,

LEWIS, W. V. (1931) The effect of wave incidence on the configuration of a shingle beach. *Geogr. J.* 78 (2), 129–48.

LIGHTHILL, M. H. and WHITHAM, G. B. (1955) On kinematic waves, I. Flood movement in long rivers. *Proc. Roy. Soc. Ser. A.* 229, 281–316.

LLIBOUTRY, L. (1958) La dynamique de la Mer de Glace et la Vaque de 1891-95,

d'aprés les mesure de Joseph Vallot. *Internat. Assoc. Sc. Hydrology Publ.* 47, 125–38 (Chamonix symposium).
LUSTIG, L. K. and BUSCH, R. D. (1967) Sediment transport in Cache Creek drainage basin in the coast ranges, west of Sacramento, California. *US Geol. Surv. Prof. Pap.* 562-A, 3, 36.
MANDELBROOT, B. B. and WALLIS, J. R. (1968) Noah, Joseph and operational hydrology. *Water Resources Res.* 4 (3), 909–18.
MARR, J. E. (1900) *The scientific study of scenery.* London. (4th edn 1912).
MASSAU, J. (1889) Appendice au Mémoire sur l'integration graphique. *Ann. Assoc. Ingénieurs sortis Ecoles Spéciales Gaud.* Ghent. 12, 185–444.
MATALAS, N. C. (1969) *Statistical design of data collection systems.* Am. Soc. Civ. Eng. Hydraulics Div. 17th Ann. Speciality Conf. Aug. 1969. 38.
MATTHEWS, J. A. (1974a) Lichen growth on an active medial moraine. Jotunheimen, Norway. *J. Glaciol.* 12, 305–13.
MATTHEWS, J. A. (1974b) Families of lichenometric dating curves from the Storbreen gletschervorfeld, Jotenheimen, Norway. *Norsk. geogr. Tidsskr.* 28, 215–35.
McCANN, S. B., HOWARTH, P. J., and COGLEY, J. G. (1972) Fluvial processes in the periglacial environment. *Trans. Inst. Brit. Geogr.* 55, 69–82.
MELTON, M. A. (1958) Geometric properties of mature drainage systems and their representation in an E_4 phase space. *J. Geol.* 66, 35–54.
MENARD, H. W. (1961) Some rates of regional erosion. *J. Geol.* 69, 154–61.
MEYERHOFF, H. A. (1960) *Time in Literature.* Berkeley and Los Angeles.
MEYERHOFF, H. A. and HUBBELL, M. (1927-8) The erosional landforms of eastern and central Vermont. *Vermont State Geologist, 16th Ann. Rep.* 315–38.
MILLER, J. P. and WENDORF, F. (1958) Alluvial chronology of the Tesoque Valley, New Mexico. *J. Geol.* 66, 177–94.
MÖRNER, N.-A. (1969) Climatic and eustatic changes during the last 15,000 years. *Geol. en. Mijnbouw.* 48 (4), 389–99.
MÖRNER, N.-A. (1971) Eustatic changes during the last 20,000 years and a method of separating isostatic and eustatic factors in an uplifted area. *Palaeogeogr., Palaeoclimatol., Palaeoecol.* 9, 153–81.
MOTTERSHEAD, D. N. and WHITE, I. D. (1972) The lichenometric dating of glacier recession, Tunsbergdal, Southern Norway. *Geogr. Ann.* 54 (A), (2), 47–52.
NAKAMURA, K. (1960) Stratigraphic studies of the pyroclastics of Oshima Volcano, Izu deposited during the last fifteen centuries. *Univ. Tokyo Sci. Pap. Coll. Gen. Educ.* 10, 125–45.
NAKAMURA, K. (1964) Volcanic stratigraphic study of Oshima Volcano, Izu. *Earthquake Res. Inst. Bull.* 42, 649–728.
NEUMANN, A. C. (1969) Quaternary sea-level data from Bermuda. *Abstracts 8th Int. Congr. Quat. Paris.* 228–9.
NYE, J. F. (1952) The mechanics of glacier flow. *J. Glaciology* 2, 82–93.
NYE, J. F. (1958) Surges in glaciers. *Nature* 181, 1450.
NYE, J. F. (1960) The response of glaciers and ice sheets to seasonal and climatic changes. *Proc. Roy. Soc.* (A) 256, 559–84.
NYE, J. F. (1963) The response of a glacier to changes in the rate of nourishment and wastage. *Proc. Roy. Soc.* (A) 275, 87–112.
NYE, J. F. (1965) The frequency response of glaciers. *J. Glaciology* 5, 567–87, 589–607.
OLSSON, I. U. (1968) C^{14}/C^{12} ratio during the last several thousand years and the reliability of C^{14} dates. In MORRISON, R. B. and WRIGHT, H. E. Jr *Means*

of correlation of Quaternary successions. Salt Lake City. 241–52.
PATERSON, W. S. B. (1969) *The physics of glaciers*. Oxford.
PAIN, C. F. (1973) Characteristics and geomorphic effects of earthquake-initiated landslides in the Adelbert Range, Papua New Guinea. *Eng. Geol.* 6(4), 261–74.
PELTIER, L. C. (1950) The geomorphic cycle in periglacial regions as it is related to climatic geomorphology. *Ann. Assoc. Am. Geogr.* 40, 214–36.
PENCK, W. (1924) *Die Morphologiche Analyse; ein Kapital der Physikalischen Geologie*. Stuttgart. Trans. (1954) CZECH, H. and BOSWELL, K. C. *Morphological analysis of landforms*. London.
PENCK, W. (1925) Die Piedmontflächen des südlichen Schwarzwaldes. *Zeits. Gessellsch. Erdk.* 83–108.
PENCK, A. and BRUCKNER, E. (1909) *Die Alpen im Eiszeitalter*, Leipzig.
PLAYFAIR, J. (1802) *Illustrations of the Huttonian theory of the earth*. Edinburgh. 528 (repr. 1956, Urbana: Ill.).
POPPER, K. (1965) *The logic of scientific discovery*. New York.
POSER, H. (1948) Boden- und Klimaverhältnisse in Mittel- und Westeuropa während der Würmeixzeit: Bonn. *Erdkunde* 2, 53–68.
POSER, H. (1954) Die Periglazial — erscheinungen in der urngebung der gletscher des zemmgrundes. *Gottinger Geographische* 15.
POWELL, J. W. (1875) *Exploration of the Colorado river of the West (1869-72)*. Washington.
PRICE, R. J. (1969) Moraines, sandar, kames and eskers near Breidamerkurjökull, Iceland. *Trans. Inst. Brit. Geogr.* 46, 17–37.
PRIOR, D. B. and STEVENS, N. (1972) Some movement patterns of temperate mudflows: examples from northeastern Ireland. *Bull. Geol. Soc. Am.* 83, 2533–44.
PUGH, J. C. (1955) Isostatic readjustment and the theory of pediplanation. *Q. J. Geol. Soc.* 111, 361–9.
QUIMPO, R. G. and YANG, J. Y. (1970) Sampling considerations in stream discharge and temperature measurements. *Water Resources Res. Am. Geophys. Un.* 6, 1771–4.
RANA, S. A., SIMONS, D. B. and MAHMOOD, K. (1973) Analysis of sediment sorting in alluvial channels. *Proc. Am. Soc. Civ. Engr. Hydraulics Div.* 99, HY11. 1967-80.
RANKAMA, K. (ed.) (1965) *The Quaternary*. New York. (2 vols.).
RAPP, A. (1961) Recent development of mountain slopes in Kärkevagge and surroundings, Northern Scandinavia. *Geogr. Ann.* 42 (2–3), 200.
RAYNER, J. N. (1971) *An introduction to spectral analysis*. London.
REGER, R. D. and PÉWÉ, T. L. (1969) Lichenometric dating in the Central Alaska Range. In PÉWÉ, T. L. (ed.), *The periglacial environment*, Montreal. 233–47.
REICHE, B. M. (1971) Land surface form in flood hydrology. In COATES, D. R. (ed.) *Environmental geomorphology*. New York. 49–68.
RICH, J. L. (1938) Recognition and significance of multiple erosion surfaces. *Bull. Geol. Soc. Am.* 49, 1695-722.
RONCA, L. B. and ZELLER, E. J. (1965) Thermoluminescence as a function of climate and temperature. *Am. J. Sci.* 263, 416—28.
RUHE, R. V. (1954) Relations of the properties of Wisconsin loess to topography in Western Iowa. *Am. J. Sci.* 252, 663–72.
RUXTON, B. P. and McDOUGALL, I. (1967) Denudation rates in northeast Papua from Potassium-argon dating of lavas. *Am. J. Sci.* 265, 545–61.
SAVIGEAR, R. A. G. (1952) Some observations on slope development in south Wales. *Trans. Inst. Brit. Geogr.* 18, 31–51.

SCHOLL, D. W., CRAIGHEAD, F. C. and STUIVER, M. (1969) Florida submergence curve revised: its relation to coastal sedimentation rates. *Sci.* 163, 562–4.
SCHEIDEGGER, A. E. (1961) Mathematical models of slope development. *Bull. Geol. Soc. Am.* 72, 37–49.
SCHEIDEGGER, A. E. (1964) Some implications of statistical mechanics in geomorphology. *Bull. Int. Assoc. Sci. Hydrol.* 9, 12–16.
SCHEIDEGGER, A. E. (1966) Effect of map scale on stream orders. *Bull. Int. Ass. Sci. Hydrol.* 11, 56–61.
SCHEIDEGGER, A. E. (1970) *Theoretical geomorphology.* Berlin, Heidelberg and New York.
SCHEIDEGGER, A. E. and LANGBEIN, W. B. (1966) Probability concepts in geomorphology. *US Geol. Surv. Prof. Pap.* 500–C, 14.
SCHOFIELD, A. N. (1971) The possible role of a centrifuge in a regional study of Calabrian slope and coast protection. *Proc. Conf. on Natural Slopes Stability and Protection.* Naples-Cosenza. 12.
SCHULMAN, E. (1956) *Dendroclimatic changes in semi-arid America.* Tuscon: Ariz.
SCHUMM, S. A. (1954) Evolution of drainage systems and slopes in badlands at Perth Amboy, New Jersey. *Columbia Univ. Dept. Geology. Tech. Rep. 8,* O.N.R. Proj. No. NR 389-042 and *Bull. Geol. Soc. Am.* 67 (1956), 597–646.
SCHUMM, S. A. (1956) The role of creep and rainwash on the retreat of badland slopes. *Am. J. Sci.* 254, 693–706.
SCHUMM, S. A. (1963a) Disparity between present rates of denudation and orogeny. *US Geol. Surv. Prof. Pap.* 454, 13.
SCHUMM, S. A. (1963b) Sinuosity of alluvial rivers on the Great Plains *Bull. Geol. Soc.* 77, 1089-1100.
SCHUMM, S. A. (1964) Seasonal variations of erosion rates and processes on hillside slopes in western Colorado. *Zeits. fur Geom. Supp.* 5, 215–38.
SCHUMM, S. A. (1965) Quaternary palaeohydrology. In WRIGHT, H. E. and FREY, D. G. (eds) *The Quaternary of the United States.* Princeton. 783–94.
SCHUMM, S. A. (1968) River adjustment to altered hydrologic regimen: Murrumbidgee River and palaeochannels Australia. *US Geol. Surv. Prof. Pap.* 598, 68.
SCHUMM, S. A. and LICHTY, R. W. (1965) Time, space and causality in geomorphology. *Am. J. Sci.* 263, 110–19.
SCHUMM, S. A. and LUSBY, G. C. (1963) Seasonal variation in infiltration capacity and runoff on hillslopes in Western Colorado. *J. Geophys. Res.* 68 (12), 3655–66.
SCHUSTER, A. (1897) On lunar and solar periodicities of earthquakes. *Proc. Roy. Soc.* 61, 455–65.
SCHUSTER, A. (1898) On the investigation of hidden periodicities with application to a supposed 26-day period of meteorological phenomena. *Terr. Magn. Atmos. Elect.* 3, 13–17.
SCHUSTER, A. (1900) The periodogram of magnetic declination. *Trans. Camb. Phil. Soc.* 18, 107–35.
SCHUSTER, A. (1906) The periodogram and its optical analogy. *Proc. Roy. Soc.* 77, 136–40.
SCROPE, G. P. T. (1858) *The geology and extinct volcanoes of central France.* London. 258 (quotation from 208–9).
SEDDON, J. A. (1900) River hydraulics. *Trans. Amer. Soc. Civil Engr.* 43, 179–243.
SELBY, M. J. (1970) *Slopes and slope processes.* Waikato Branch New Zealand

Geog. Soc. Publ. No. 1.

SHAW, A. B. (1964) *Time in stratigraphy*, New York.

SHEPARD, F. P. (1963) Thirty-five thousand years of sea level. In CLEMENTS, T. (ed.) *Essays in marine geology in honour of K.O. Emery*. Los Angeles. 1–10.

SHERMAN, L. K. (1932) Streamflow from rainfall by unit-graph method. *Engr. News Record* 108, 501–5.

SIGAFOOS, R. S. and HENDRICKS, E. L. (1961) Botanical evidence of the modern history of the Nisqually Glacier, Washington. *US Geol. Surv. Prof. Pap.* 387-A, 20.

SIMONETT, D. S. and ROGERS, D. L. (1970) The contribution of landslides to regional denudation in New Guinea. *US Office Naval Res. Tech. Rep.* 6. Geog. Branch Contract O.N.R. 583-(11) Task No. NR 389-133. Lawrence: Kansas. 34.

SLAYMAKER, O. (1972) Patterns of present sub-aerial erosion and landforms in mid-Wales. *Trans. Inst. Brit. Geogr.* 55, 47–68.

SMITH, T. R. and BRETHERTON, F. P. (1972) Stability and the conservation of mass in drainage basin evolution. *Water Resources Res.* 8, (6), 1506–27.

SOERGEL, W. (1924) *Die Diluvialen Terrassen der Ilm und ihre Bedeutung für die Gliederung des Eiszeitalters.* Jena.

SPARKS, B. W. (1953) The former occurrence of both Helicella striata (Müller) and H. geyeri (Soós) in England. *J. Conch.* 23, 372–8.

STEERS, J. A. (1948) *The coastline of England and Wales.* Cambridge.

STODDART, D. R. (1966) Darwin's impact on geography. *Ann. Ass. Am. Geogr.* 56 (4), 683–98.

STODDART, D. R. (1969) Climatic geomorphology: review and re-assessment. In BOARD, C. *et al.* (eds) *Progress in Geography* 1, 161–222.

STONE, R. (1961) Geologic and engineering significance of changes in elevation revealed by precise levelling, Los Angeles area, California. *Geol. Soc. Am. Spec. Pap.* 68, 57–8.

STORK, A. (1963) Plant immigration in front of retreating glaciers, with examples from the Kebnekajse area, northern Sweden. *Geog. Ann.* 45, (1), 1–22.

STRAHLER, A. N. (1950) Equilibrium theory of erosional slopes approached by frequency distribution analysis. *Am. J. Sci.* 248, 673–96, 800–14.

STRAHLER, A. N. (1951) *Physical geography.* New York. (3rd edn 1969).

STRAHLER, A. N. (1952) Dynamic basis of geomorphology. *Bull. Geol. Soc. Am.* 63, 923–38.

STRAKHOV, A. N. (1967) *Principles of lithogenesis.* Trans. FITZSIMMONS, J. P., ed. TOMKEIFF, S. I. and HEMINGWAY, J. E. New York and Edinburgh. (3 vols.).

STRAW, A. B. (1964) *Time in stratigraphy.* Internat. Series in the Earth Sciences, New York.

SUBCOMMITTEE ON SEDIMENTATION (1953) Summary of reservoir sedimentation surveys of the United States through 1950. Federal Inter-Agency River Basin Comm. *Sedimentation Bull.* 5, 222.

SUGGATE, R. P. (1968) Post-glacial sea-level rise in the Christchurch Metropolitan area, New Zealand. *Geol. en Mijnbouw* 47, 291–7.

TASK COMMITTEE ON PREPARATION OF SEDIMENTATION MANUAL, AMERICAN SOCIETY OF CIVIL ENGINEERING (1970) Sediment measurement techniques. Ch. III, Reservoir Deposits. *Proc. Am. Soc. Civil Engr. Hydraulics Div.* 96, HY12, 2417–46.

THOMPSON, J. R. (1964) Quantitative effect of watershed variables on rate of gully-head advancement. *Trans. Am. Agr. Engr.* 7 (1), 54–5.

THORNES, J. B. (1967) *Erosion and Sedimentation in the Alto Duero, Spain* Unpubl. Ph.D. thesis, University of London.

THORNES, J. B. (1968) A queuing theory analogue for scree slope studies. *Graduate School of Geography, London School of Economics Disc. Paper* 22.

THORNES, J. B. (1971a) Rivers in their delicate courses. *Geogr. Mag.* Dec. 1971, 5.

THORNES, J. B. (1971b) State, environment and attribute in scree-slope studies. In BRUNSDEN, D. (comp.) *Slopes, form and process.* Inst. Brit. Geogr. Spec. Publ. 3, 49–63.

THORNES, J. B. (1972) Debris slopes as series. *J. Arctic and Alpine Res.* 4, 337–42.

THORNES, J. B. (1973) Markov chains and slope series: the scale problem. *Geogr. Ann.* 5 (4), 322–8.

TRICART, J. (1956) Études expérimentale du problème de la gelivation. *Biul. Peryglac.* 4, 285–318.

TROLL, C. (1943) Die Frostwechselhaufigkeit in den Tuft — und Bodenklunaten der Eide. *Meteorol. Zeits.* 60, 161–71.

TROLL, C. (1944) Strukturboden, Solifluction und Frostklimate der Erde. *Geologische Rundschau* 34 (7/8), 545–694.

TSUBOI, C. (1933) Investigation on the deformation of the earth's crust found by precise geodetic means. *Japanese J. Astron. and Geophys.* 10, 93–248.

TWIDALE, C. R. (1968) *Geomorphology: with special reference to Australia.* Melbourne.

VARNES, D. J. (1958) Landslide types and processes. 20–47. In *Landslides and engineering practice* (ed. ECKEL, E. B.). Highway. Res. Board. Spec. Rep. 29.

VITA-FINZI, C. (1969) *The Mediterranean valleys: geological changes in historical times.* Cambridge.

VON ENGELN, O. D. (1942) *Geomorphology.* New York (5th edn 1956).

VON SCHELLING, T. (1951) Most frequent particle paths in a plane. *Trans. Am. Geophy. Un.* 32, 222.

VOSS, J. (1933) Pleistocene forests of central Illinois. *Bot. Gaz.* 94, 808–14.

WEISS, L. L. (1955) A nomogram based on the theory of extreme values for determining values for various return periods. *Monthly Weath. Rev.* Mar. 69–71.

WELCH, D. M. (1970) Substitution of space for time in a study of slope development. *J. Geol.* 78, 234–8.

WEYMAN, D. R. (1970) Throughflow on hillslopes and its relation to the stream hydrograph. *Bull. Int. Assoc. Scient. Hydrol.* 15, 25–33.

WHIPKEY, R. Z. (1965) Subsurface stormflow from forested slopes. *Bull. Int. Assoc. Scient. Hydrol.* 10, 74–85.

WHITAKER, W. (1867a) On sub aerial denudation, and on cliffs and escarpments of the Chalk and Lower Tertiary beds. *Geol. Mag.* 4, 447–54, 483–93.

WHITAKER, W. (1867b) On subaerial denudation, and on cliffs and escarpments of the Tertiary strata. *Q. J. Geol. Soc.* 23, 265–6.

WHITAKER, W. (1868) Subaerial denudation. *Geol. Mag.* 5, 46–7.

WILLIAMS, G. P. (1967) Flume experiments, transport of coarse sand. *US Geol. Surv. Prof. Pap.* 562–B, 31.

WILSON, A. G. (1970) *Entropy in urban and regional modelling.* London.

WIMAN, S. (1963) A preliminary study of experimental frost weathering. *Geogr. Ann.* 45, 113–21.

WOLDENBURG, M. J. (1966) Horton's laws justified in terms of allometric growth. *Bull. Geol. Soc. Am.* 77, 431–4.

WOLF, P. O. (1966) Comparison of methods of flood estimation. *Proc. Inst. Civ. Engrs. Symp. on River Flood Hydrol.* 1–23.

WOLF, P. O. (1966) Notes on the management of water resource systems. *J. Inst. Water Engrs.* 20 (2), 95–105.

WOOD, A. (1942) The development of hillside slopes. *Proc. Geol. Assoc.* 53, 128–40.

WOOLDRIDGE, S. W. (1948) The role and relations of geomorphology. Inaugural Lecture, King's College London, repr. in *The geographer as scientist* (1956). London.

WOOLDRIDGE, S. W. and LINTON, D. L. (1955) *Structure, surface and drainage in south-east England.* London. (1st edn 1939.)

WOOLDRIDGE, S. W. and MORGAN, R. S. (1937) *The physical basis of geography: an outline of geomorphology.* London.

WRIGHT, G. F. (1881) An attempt to calculate approximately the date of the glacial era in eastern North America. *Am. J. Sci.* 21, 120–3.

YANG, C. T. (1970) On river meanders. *J. Hydrol.* 13, 231–53.

YANG, C. T. (1971) Potential energy and stream morphology. *Water Resources Res.* 7(2), 311–22.

YASSO, W. E. (1971) Forms and cycles in beach erosion and deposition. In COATES, D. R. (ed.) *Environmental geomorphology*, New York, 109–37.

YOUNG, A. (1963) Deductive models of slope evolution. *I.G.U. Slope Comm. Report* 3, 45–66. Also in *Neue Beiträge zur internationalen Hangforschung.* Gottingen.

YOUNG, A. (1969) Present rate of land erosion. *Nature* 224, 851–2.

YOUNG, A. (1972) *Slopes.* London.

ZEUNER, F. E. (1945) *The Pleistocene period: its climate, chronology and faunal successions.* London.

ZEUNER, F. E. (1952) *Dating the past: an introduction to geochronology* London.

Appendix

The metric system: conversion factors and symbols

In common with several other text book series *The Field of Geography* uses the metric units of measurement recommended for scientific journals by the Royal Society Conference of Editors.* For geography texts the most commonly used of these units are:

Physical quantity	*Name of unit*	*Symbol for unit*	*Definition of non-basic units*
length	metre	m	basic
area	square metre	m^2	basic
	hectare	ha	$10^4 m^2$
mass	kilogramme	kg	basic
	tonne	t	$10^3 kg$
volume	cubic metre	m^3	basic-derived
	litre	l	$10^{-3} m^3$, 1 dm^3
time	second	s	basic
force	newton	N	kg m s^{-2}
pressure	bar	bar	$10^5 Nm^{-2}$
energy	joule	J	kgm^2s^{-2}
power	watt	W	$kgm^2s^{-3} = Js^{-1}$
thermodynamic temperature	degree Kelvin	°K	basic
customary temperature, *t*	degree Celsius	°C	$t/°C = T/°K - 273.15$

Fractions and multiples

Fraction	*Prefix*	*Symbol*	*Multiple*	*Prefix*	*Symbol*
10^{-1}	deci	d	10	deka	da
10^{-2}	centi	c	10^2	hecto	h
10^{-3}	milli	m	10^3	kilo	k
10^{-6}	micro	μ	10^6	mega	M

The gramme (g) is used in association with numerical prefixes to avoid such absurdities as mkg for μg or kkg for Mg.

Conversion of common British units to metric units

Length

1 mile	= 1.609 km	1 fathom	= 1.829 m
1 furlong	= 0.201 km	1 yard	= 0.914 m
1 chain	= 20.117 m	1 foot	= 0.305 m
		1 inch	= 25.4 mm

Area

1 sq mile	= 2.590 km^2	1 sq foot	= 0.093 m^2
1 acre	= 0.405 ha	1 sq inch	= 645.16 mm^2

Mass

1 ton	= 1.016 t	1 lb	= 0.454 kg
1 cwt	= 50.802 kg	1 oz	= 28.350 g
1 stone	= 6.350 kg		

Mass per unit length and per unit area

1 ton/mile	= 0.631 t/km	1 ton/sq mile	= 392.298 kg/km^2
1 lb/ft	= 1.488 kg/m	1 cwt/acre	= 125.535 kg/ha

Volume and capacity

1 cubic foot	= 0.028 m^3	1 gallon	= 4.546 l
1 cubic inch	= 1638.71 mm^3	1 US gallon	= 3.785 l
1 US barrel	= 0.159 m^3	1 quart	= 1.137 l
1 bushel	= 0.036 m^3	1 pint	= 0.568 l
		1 gill	= 0.142 l

Velocity

1 m.p.h.	= 1.609 km/h
1 ft/s	= 0.305 m/s
1 UK knot	= 1.853 km/h

Mass carried × distance

1 ton mile = 1.635 t km

Force

1 ton-force	= 9.964 kN
1 lb-force	= 4.448 n
1 poundal	= 0.138 N
1 dyn	= 10^{5}

Pressure

1 ton-force/ft^2	= 107.252 kN/m^2
1 lb-force/in^2	= 68.948 mbar
1 pdl/ft^2	= 1.488 N/m^2

Energy and power

1 therm	= 105.506 MJ
1 hp	= 745.700 W(J/s) = 0.746 kW
1 hp/hour	= 2.685 MJ
1 kWh	= 3.6 MJ
1 Btu	= 1.055 kJ
1 ft lb-force	= .356 J
1 ft pdl	= 0.042 J
1 cal	= 4.187 J
1 erg	= 10^{-7} J

Metric units have been used in the text wherever possible. British or other standard equivalents have been added in brackets in a few cases where metric units are still only used infrequently by English-speaking readers.

* Royal Society Conference of Editors, *Metrication in Scientific Journals*, London, 1968.

Index

Author Index

Ahnert, F. 129, 186
Anderson, H. W. 18, 186
Anderson, J. L. 49, 50, 186
Andrews, J. T. 49, 80, 186
Antevs, E. 48, 186
Arber, M. A. 54, 186
Bagnold, R. A. 114, 186
Bakker, J. P. 145, 186
Beckinsale, R. P. 19, 20, 120, 188
Benedict, J. B. 49, 50, 186
Benson, M. A. 65, 192
Berry, B. J. L. 182-3, 186
Beschel, R. 49, 186
Birman, J. H. 39, 187
Birot, P. 125, 187
Black, R. D. 103, 189
Bloom, A. L. 181, 187
Boer, G. de 55, 187
Bolton, G. S. 39, 187
Bonham-Carter, G. 166, 191
Bornfeldt, F. 50, 187
Bray, D. J. 102, 192
Bretherton, F. P. 153-4, 197
Brillouin, L. 173, 187
Brown, E. H. 109, 115, 123, 187
Brown, J. C. 34, 187
Bruckner, E. 42, 195
Brunsden, D. 11, 14, 16, 18, 24, 59, 105, 119, 131, 187
Bryan, K. 56, 123, 187
Buckland, W. 53, 187
Busch, R. D. 109, 194
Butzer, K. W. 127-8, 187
Calver, A. 95, 187
Carr, A. P. 56, 187
Carson, M. A. 21, 93, 94, 103, 111, 115, 121, 147, 187
Carter, C. S. 24, 188
Chamberlin, T. C. 38, 188
Chandrasekhar, S. 169, 188
Chorley, R. J. 10, 19, 20, 24, 112, 113, 115, 120, 121, 132, 171, 188
Churchward, H. M. 38, 188
Clayton, K. M. 32, 188
Cogley, J. G. 57, 60, 194
Colby, B. R. 102, 188
Conybeare, W. D. 53, 188
Cook, F. A. 60, 61, 188
Cooper, H. H. 142, 188
Corbel, J. 93, 110, 188
Craighead, F. C. 79, 195
Culling, W. E. H. 3, 23, 139, 140, 157, 169-71, 188
Darwin, C. R. 56, 188
Davis, M. B. 33, 188
Davis, W. M. 5, 114, 115, 120-3, 125, 132, 155, 188-9
Derbyshire, E. 124, 189
Dole, R. B. 94, 189
Douglass, A. E. 44, 189
Dunn, A. J. 19, 20, 120, 188
Dunne, T. 103, 104, 189
Dury, G. H. 107, 189
Dutton, C. E. 56, 189
Eagleson, P. 146, 189
Edwards, A. M. C. 131, 189
Eicher, D. L. 41, 189
Elias, F. 77, 189
Emiliani, C. 35, 189-90
Emmett, W. W. 93, 100, 105, 110, 190, 193
Evans, I. 136, 190

Evans, P. 35, 190
Fairbridge, R. W. 109, 190
Fisher, O. 145, 190
Flaxman, E. M. 94, 190
Fleming, G. 108, 190
Flint, R. F. 42, 190
Fourier, J. B. 81–4, 190
Fournier, F. 94, 108, 190
Freeman, J. R. 18, 190
Frye, J. C. 31, 190
Gardner, J. 59, 190
Garner, H. F. 125, 190
Geer, G. de 48, 190
Gilbert, G. K. 5, 114, 118, 129, 190
Gilloly, J. 94, 190
Goodlett, J. C. 33, 188
Graybill, F. A. 98, 193
Green, J. F. N. 30, 190
Gregory, K. J. 26, 101, 102, 190
Grodins, F. S. 140, 191
Gudgin, J. 177, 191
Gumbel, E. J. 75, 191
Gupta, V. L. 66, 191
Guthrie-Smith, H. 56, 191
Gutenberg, B. 94, 191
Hack, J. T. 118, 173, 191
Hall, E. T. 49, 191
Harbaugh, J. W. 166, 191
Harvey, D. 5, 112–115, 191
Hendricks, A. L. 45, 46, 197
Hewitt, K. 57, 75, 191
High, R. D. 94, 190
Hilton, P. 136, 191
Hirano, M. 139, 191
Holeman, J. N. 94, 191
Horton, R. E. 150, 191
Howard, A. D. 3, 159, 191
Howarth, P. J. 57, 60, 194
Hubbell, D. W. 102, 191
Hurst, H. E. 56, 191
Hutchinson, J. T. 119, 191
Hutton, J. 5, 191
Jackli, H. 110, 192
James, P. A. 63, 192
Jelgersma, S. 79, 192
Jenkins, G. M. 87, 192
Jennings, M. E. 65, 192
Jochimsen, M. 49, 192
Johnson, D. W. 115, 132, 192
Johnson, R. H. 33, 192
Jones, D. K. C. 14, 34, 123, 131, 187, 192
Jones, O. T. 30, 192
Kellerhals, R. 102, 192
Kennedy, B. 10, 24, 113, 188
Kerney, M. P. 36, 37, 192
Kesel, R. H. 16, 24, 119, 187
Kidson, C. 54, 192
King, C. A. M. 25, 99, 129, 161, 192
King, L. C. 30, 114, 117, 121, 192–3
Kirkby, M. J. 5, 93, 94, 95, 103, 111, 114, 121, 129, 147, 152–3, 187, 193
Knapp, B. H. 64, 103, 193
Krumbein, W. C. 41, 69, 98, 193
Larsson, W. 40, 193
La Marche, V. C. Jn. 45, 193
Langbein, W. B. 108, 149, 155, 164, 171, 193
Lees, G. M. 94, 193
Le Heux, J. W. B. 145, 186
Lehmann, O. 145, 147, 193
Leighly, J. 5, 193
Leonard, A. B. 31, 190
Leopold, L. B. 38, 56, 76, 93, 104, 105, 110, 149, 163, 171, 193
Lewis, W. V. 161, 193
Lichty, R. W. 13, 15, 68, 196
Lighthill, M. H. 149, 193
Linton, D. L. 20, 31, 123, 199
Lliboutry, L. 151, 193
Lustig, L. 109, 194
Mahmood, K. 95, 195
Mandelbroot, B. B. 84, 194
Massau, J. 149, 194
Matalas, B. N. 66, 194
Matthews, J. A. 49, 50, 51, 194
McCullagh, M. J. 25, 161, 192
McCann, S. B. 57, 60, 194
McDougall, I. 107, 195
Melton, M. A. 113, 155, 194
Menard, H. W. 106, 109, 194
Meyerhoff, H. A. 4, 194
Miller, J. P. 41, 106, 194
Mottershead, N. B. 49, 50, 194
Mörner, N-A. 78, 79, 194
Myrick, R. M. 93, 105, 110, 193
Nakamura, K. 39, 194
Neumann, A. C. 79, 194
Nye, J. F. 3, 129, 150, 151, 194
Olsson, I. U. 43, 194
Österberg, M. 50, 187
Paterson, W. S. B. 148, 194
Pain, C. F. 119, 195
Peltier, L. C. 124, 125, 132, 195
Penck, A. 42, 195
Penck, W. 114, 121–3, 195
Petley, D. J. 21, 115, 121, 187
Péwé, T. L. 49, 195

Popper, K. 23, 195
Poser, H. 111, 195
Powell, J. W. 56, 123, 195
Price, R. J. 131, 195
Prior, D. B. 16, 81, 195
Pugh, J. C. 119, 195
Quimpo, R. G. 66, 195
Raiche, V. G. 60, 61, 188
Rana, S. A. 95, 195
Rankama, K. 47, 195
Rapp, A. 93, 110, 195
Rayner, J. N. 87, 195
Reger, R. D. 49, 195
Reiche, B. M. 9, 195
Rogers, D. L. 24, 197
Ronca, L. B. 49, 195
Rorabaugh, M. J. 142, 188
Ruhe, R. V. 107, 195
Ruxton, B. P. 107, 195
Savigear, R. A. G. 21, 24, 195
Scholl, D. W. 79, 195
Scheidegger, A. E. 137–40, 148, 155, 157, 166, 170–1, 173, 196
Schofield, A. N. 100, 196
Schumm, S. A. 5, 13, 15, 24, 26, 68, 93, 94, 105, 108, 111, 113, 121, 193, 196
Schulman, E. 196
Schuster, A. 83, 196
Scrope, G. P. T. 1, 196
Seddon, J. A. 149, 196
Selby, M. J. 196
Shaw, A. B. 35, 197
Shepard, F. P. 79, 197
Sherman, L. K. 90, 197
Sigafoos, R. S. 45, 46, 197
Simonett, D. S. 24, 197
Simons, D. B. 95, 195
Slaymaker, H. O. 111, 197
Sloss, L. L. 41, 193
Smith, T. R. 154, 197
Soergel, W. 127, 197
Sollid, J. L. 49, 50, 186
Sparks, B. W. 36, 197
Stabler, H. 94, 189
Steers, J. A. 55, 197
Stevens, N. 16, 81, 195
Stoddart, D. R. 5, 94, 120, 124, 195
Stone, R. 94, 197
Stork, A. 49, 197
Strahler, A. N. 10, 24, 113, 169, 197
Strakhov, A. N. 108, 197
Straw, A. 41, 197
Stuiver, M. 79, 195
Subcommittee on Sedimentation 106, 197
Suggate, R. P. 79, 197
Thompson, J. R. 94, 197
Thornes, J. B. 17, 39, 69, 73, 86, 131, 159, 167, 177, 178, 197, 198
Tricart, J. 100, 198
Troll, C. 93, 198
Tsuboi, C. 94, 198
Twidale, C. R. 119, 198
Varnes, D. J. 98, 99, 198
Vita-Finzi, C. 29, 198
Von Engeln, O. D. 122, 198
Von Schelling, T. 157, 198
Voss, J. 38, 198
Walling, D. E. 26, 101, 102, 190
Wallis, J. R. 84, 194
Watts, D. G. 87, 192
Webber, P. 49, 186
Weiss, L. L. 78, 198
Welch, D. M. 24, 119, 198
Wendorf, F. 41, 106, 194
Weyman, D. 95, 103, 187
Whipkey, R. Z. 103, 194
Whitham, G. B. 149, 193
White, I. D. 49, 50, 194
Williams, G. P. 101, 198
Wilson, A. G. 173, 198
Wiman, S. 100, 198
Woldenburg, M. J. 113, 198
Wolf, P. O. 65, 198–9
Wolman, M. G. 193
Wood, A. 122, 199
Wooldridge, S. W. 20, 31, 120, 123, 199
Worsley, P. 39, 187
Wright, G. F. 42, 199
Yang, C. T. 171, 173, 199
Yang, J. Y. 66, 195
Young, A. 21, 22, 103, 199
Zeuner, F. E. 41, 117, 198
Zeller, E. J. 49, 195

Subject Index

References to figures are given in italics

absolute dating: dendrochronology 44–7; *2.9, 2.10*; historical records 50–2; lichenometry 49–50, *2.12, 2.13*; radiometric techniques 42–4; thermoluminescence 48–9; varves 47–8, *2.11*
acceleration 3
aerial photographs 52–6

analysis of temporal data 68–92; discrete serially independent events 70–8, *4.2, 4.3, 4.4, 4.5, 4.6,* tables 4.1, 4.2; serially dependent events 78–92; periodicity 81–4, *4.9, 4.10*; persistence 84–7, *4.11, 4.12*; transfer function approach 89–92, *4.15, 4.16*; trend 78–81, *4.7, 4.8*; two or more series 87–9, *4.13, 4.14*
antecedent conditions 13
archaeological dating 41
artifacts 40–1
autocorrelation 69, 70, 84
autocovariance 87
autoregressive: model 87; process 19, 84, 87

base level *6.2*
Bernoulli process 72
binominal process 72

calculus 135–7, *7.1, 7.2*
Cartesian coordinates *9.1*
cascaded systems 13, *1.10*
characteristic form of hillslopes 152–3, *7.13*
climatic curves, uses for dating 35, *2.5*
climate and denudation rates 93–4, 108–11, *5.5, 6.4, 6.5,* table 5.2
climatic geomorphology 20, 124–9, *6.4, 6.5*
coherence 87–9, *4.14*
continuity principle and equations 142–9; steady state equation of 142–5; applied to glaciers 146, *7.12*; applied to overland flow 146; applied to hillslope erosion 147–53, *7.99, 7.13*
continuous time 6, *1.4, 1.8*
correlogram 70, 84, *4.11*
cross-correlation 87
cumulative exponential distribution 74, *4.4*
cumulative frequency 9, *1.6*
cumulative probability distribution 70, *4.2*
cycles 62, *3.5*
cyclic models 119–23, *6.2*
cyclic time 15, Table 1.1

data: classification of 3; serially independent and dependent 68–70; stationary 68–70
dating 28–52
Davisian model 114–15, 117, 120–3, *6.2*; criticisms of 121; interruptions to 120
decay models 119–23; of vertical cliff 139, *7.4*; two dimensional deterministic 148–9, *7.11*
dendrochronology 44–7, *2.9, 2.10*
denudation chronology 20, 121–4, table 1.2; and Davisian model 120–1; model for south-east England 123–4, *6.3*
denudation rates 108–11, *5.5,* tables 5.1b and 5.2; specific to environment 110–11; specific processes 111
differentiation 135–6
differential equations 136–42; partial 136–7; non linear 139
diffusion 142–5; model *7.7*; probabilistic development of 169–71
digital simulation 158–62
discrete serially independent events 70–8, *4.2, 4.3, 4.5, 4.6,* tables 4.1, 4.2
discrete time 7, *1.4*
documentary evidence 56
drainage patterns 31–2
dynamic equilibrium 118–19

entropy: general principles 171–3; maximization 173; Clausian 171; Shannonian 172; river meanders 173
equilibrium models 118–19; process specific 129; climatic control 124–5
equifinality 1, 22
erosion surfaces 20, 29–30, *2.1*; isochronous and metachronous 30, *2.2*
ergodic transformation 23–5, 164, 177
events: catastrophic 53–5, 57, *3.1, 3.3*; continuous in time 57; irregular in time 57; non events 65; observation of 56–61; recording of 63–4; regular 56; seasonal 57–9, *3.4,* table 3.2
evolution 5, 19; of slopes 21–4, 131, *1.15*
extreme events 75–8
explanation 4

feedback 11, 114, *1.9*; negative 118–20; positive 129–30
filter 19, 86, 92
fixed time models 117, 126–8; qualitative, global 123; qualitative, with process assemblages 126–8; qualitative, process specific 132

floating chronology models 117–18, 124–5
forecasting, general neglect of 23, 26
fossils: Foraminifera 35–6; molluscs 36, *2.6*
Fourier analysis 81–3, *4.9*
Fourier transform 84
frequency 7–10, 70, *1.3*
frequency and magnitude 8–10, 75

Gamma process 74
geochronology, principles of 41–52
global models, fixed scale 123
Gumbel distribution 77, *4.6*, Table 4.2

historical records 50–6, *3.1, 3.2, 3.3*
hydrograph 65; unit 90–1, *4.16*

impulse response function 91–2
information content 66–7, *3.8*
input-output 11, 19, 134–5, *1.9, 1.10*
integration 7
isochronous surfaces 30, *2.2*

kinematic waves 3, 149; meaning of 149; application of in glaciers 150–1; application of in overland flow 150
King's model of landform evolution 121–3, *6.2*

lags in systems 10–19, *1.8*
landscape, reduction of 15
lichenometry 49–50, *2.11, 2.13*

magnitude and frequency 9, 75
manifolds 182–3, *9.5*
map evidence, in dating 55–6
Markov chain 25, 157, 164–6
Markov process 72, 166–7
measurement: morphological techniques 28–9; ordinal scale 28; ratio scale 28; types of 58, *Table 3.1, 5.1*
memory, of a system 19
metachronous erosion surfaces 30, *2.2*
models: analogue 112; calibration of 23; classification of according to function, 112–15; evaluation of 115–16; steps in construction of 138; validation of 19; verification of 23
models and theory 113–15
Monte Carlo techniques 157–8, 162
moving average process 19, 85

negative feedback 118–20
noise 85

observation of variables in time: controlled 56–61; experimental design 61–3; incidental 53–6; recording 63–4; recurrence 64–5; sub-optimal 56
ocean cores 35, *2.5*
ordinal scale 28

palaeosols 36–9, *2.7*
palynology 32–5, *2.3, 2.4*
partial duration series 76
Pearson type III distribution 78
Penck's model of landform evolution 121–3, *6.2*
periodicity 81–4, *4.9, 4.10*
persistence 65, 84–7, *4.11, 4.12*
Poisson process 72–3
pollen analysis *see* palynology
positive feedback 129–30
probability distributions 72–8
probability density function 70, *4.2*
process-specific models 129–32

qualitative global models 123
quantization 6, *1.2, 1.4*
queues 73–4

radiocarbon dating 42–3
radionuclides 42–4
randomness 3
random walk 162–4, *8.2*
rates of operation of processes: application to engineering and management 96; affected by man 23; climates in relation to 93–4, *5.5*; experiments on 98; field 101–6, *5.2, 5.3*; general proposition of measurement 15, 96–101; inferred 106–8, *5.4*; reasons for investigation of 93; sampling of 59–60; specific rates of denudation 108–11; types of information 98–9
ratio scale 28
recording techniques 63–4, *3.6, 3.7*
recurrence 64, 74–8, *4.5*
relative dating: artifacts 40–1; drainage patterns 31–2; erosion surfaces 29–30; organic remains 32–6, *2.4*; sea levels 30–1; stratigraphical 32; weathered products and relict deposits 36–41
relative time models 125–6
relaxation: path 16–20, *1.14*; time 16, 113, 114
resistance to change 17–19

response: function, impulse 91–2; function, non-linear 92; dynamic, to a threshold 16, *1.12*
resultant 3–4
return period 9, *1.7*

sampled time 7, *1.4*
sampling 6–7, 57–67, *3.4, 3.5*, table 3.1
scale: relative time and space 183–5
sea levels: use of, as basis of correlation 30–1; height of 30–1
sediment: bedload sampling 102–3; in reservoirs 106–7; suspended 109
serial dependence 70–8
serially dependent data 68–70
serially dependent events 78
series: partial duration 76–7, *4.6, Table 4.2*; spatial 24
simulation 25, 157–62, *1.16*; basic rationale for 158; types of 26; of a spit 161–2; of stream capture 159–61; problems of 161
slopes: and ergodic systems 24; evolution of 21
soil creep, Culling's model 23, 169
space and time 3, 14, 174–84; basic structures of 175–7, table 9.1; differences between 174; manifolds 182–3, *9.5*; models of 174–85; model of sea-level change *9.4*; transformation 22–5, types of data table 9.1
spatial series 24
spatial variability 15, 23, *1.11*
spectrum 83
state variable 12
stationarity 69
statistics: extremal 75–8; indicators of stability 15; information content of 66–7, *3.8*
steady state 142–5
stochastic models: autoregressive 87; basic types 156; continuous time and space states 168–71; Monte Carlo 157–8; processes 72–4, 162–8; random walk 162–4; rationale for 155–6; of long profile 162–3, *8.2*; of soil creep 168–70; of stream order 166
stores: as queues 167; in systems 11, 26
sulphurous smells 53
system: cascaded 13, *1.10*; constraints 19; definition 10; equations of 140–1; ergodic 23–5, 164; geomorphological 10–19; memory 13, 19, 69; non-anticipatory 11; order 140–2, *7.5, 7.6*; output of 15; relaxation in 16; shock to 15, 80; stability of 15; states in 12; structure of 10; thresholds in 15

temporal models 112–15
tephrachronology 39–40, *2.8*
terraces 29, 30, 127
thermoluminescence 48
thresholds 15, 64
time: absolution position 41–52; and space 174–85; centrality in geomorphology 2; cyclic 15, table 1.1; description of events in 5–10; evolution in 5; explanation 2–5; graded 15, table 1.1; independence of 117–19; measurement of 28–41; relative position in 28–41; status of variables in 13–16, *1.1*; steady 15, *1.1*
time series 78–92; harmonic analysis of 81–3; spectral analysis of 82–4
transfer function 89-92, 114, 134, *6.1*;
trend 15, 62, 78–81, *3.5, 4.7, 4.8*
two or more series 87–9, 4.13, 4.14

unit hydrograph 90–1, *4.16*

variables in time, measurement of 53–67
varves 47–8, *2.11*
vector 3–5
velocity 3
verification, of evolutionary model 23–5

weathered products 36

For Product Safety Concerns and Information please contact our EU representative GPSR@taylorandfrancis.com
Taylor & Francis Verlag GmbH, Kaufingerstraße 24, 80331 München, Germany

www.ingramcontent.com/pod-product-compliance
Ingram Content Group UK Ltd.
Pitfield, Milton Keynes, MK11 3LW, UK
UKHW020201200926
475702UK00004B/556
* 9 7 8 0 3 6 7 2 2 0 0 5 1 *